Great MYTH CONCEPTIONS

Great MYTH CONCEPTIONS

Dr Karl Kruszelnicki

Illustrated by Adam Yazxhi

HarperCollins*Publishers*

HarperCollins*Publishers*

First published in Australia in 2004
by HarperCollins*Publishers* Pty Limited
ABN 36 009 913 517
A member of the HarperCollins*Publishers* (Australia) Pty Limited Group
www.harpercollins.com.au

HarperCollins*Publishers*
25 Ryde Road, Pymble, Sydney NSW 2073, Australia
31 View Road, Glenfield, Auckland 10, New Zealand
77–85 Fulham Palace Road, London W6 8JB, United Kingdom
2 Bloor Street East, 20th floor, Toronto, Ontario M4W 1A8, Canada
10 East 53rd Street, New York NY 10022, USA

National Library of Australia Cataloguing-in-publication data:

Kruszelnicki, Karl, 1948– .
Great mythconceptions : cellulite, camel humps and chocolate zits.
ISBN 0 7322 8062 1.
1. Science – Popular works. 2. Errors,
Popular. I. Yazxhi, Adam. II. Title.
500

Cover photo by Adam Craven
Internal design and layout: Judi Rowe, Agave Creative Design
Printed and bound in Australia by Griffin Press on 80gsm Fine Offset

5 4 3 2 04 05 06 07

To Ganesha, the God of overcoming obstacles.
Thanks for getting us safely out of the Himalayas.

Contents

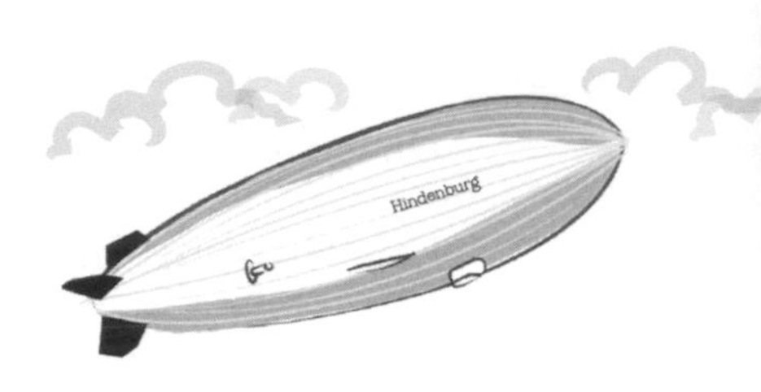
Hindenburg

Quantum

Dirty Desks

If things get a bit rushed at work, you might grab a quick sandwich at your office desk. On the other hand, you would never dream of eating off the toilet seat, because everyone 'knows' that toilets are 'dirty', and loaded with germs. But, on average, a desk has 50 times more bacteria per square centimetre than a toilet seat!

Dr Charles Gerba, a microbiologist from the University of Arizona, discovered this fact. He's 'Dr Germs'. Over the past three decades, he's written some 400 papers on infection and disinfection in peer-reviewed journals.

He solved the problems that the National Science Foundation was having with the waste-water treatment system in the Antarctic at McMurdo Station. He helped out with advice on water-recycling systems for both NASA and the Russian *Mir* Space Station. He loves his work so much, that he even gave his first son the middle name of Escherichia, which is the 'E' in *E. coli*, the famous faecal bacterium. He got around family resistance by telling his father-in-law that Escherichia was the name of a king in the Old Testament of the Bible.

From June to August of 2001, he and his team looked for five different types of bacteria — *E. coli, Klebsiella pneumonia, Streptococcus, Salmonella* and *Staphylococcus aureus*. The team studied offices at four United States locations — New York City, San Francisco, Tampa and Tucson. At each site, they tested surfaces three times a day for five days. They sampled

12 different surfaces — desktop, telephone receiver, computer mouse, computer keyboard, microwave door button, elevator button, photocopier start button, photocopier surface, toilet seat, fax machine, refrigerator door handle and water fountain handle. The team wanted to measure the effect of cleaning each surface. At each location, one group of employees used disinfecting wipes to clean the surfaces they worked with, while the other group did not. (The study was partly funded by Clorox, a company that makes disinfecting wipes.)

The results were astonishing. In terms of bacteria per square inch (25.4 mm^2), they found that the telephone receiver was the filthiest — 25 127 (probably because many people often share the same phone). This was followed by the desktop at 20 961, the computer keyboard at 3295 and the computer mouse at 1676. The least contaminated surface was the toilet seat with only 49 bacteria per square inch — making it about 400 times cleaner than the desktop. Gerba says that, for bacteria, the 'desk is really

Dirty Rotten Desks

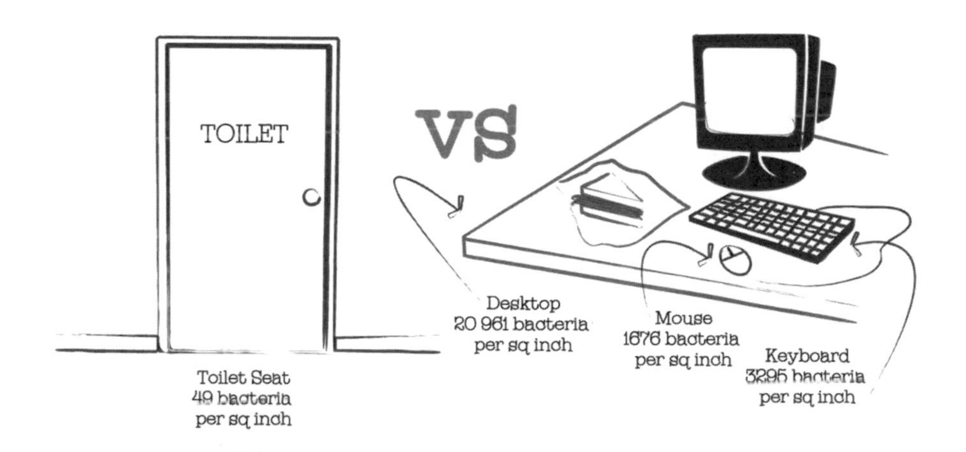

Dirty Desks

In terms of bacteria per square inch (25.4 mm^2), the telephone receiver was the filthiest followed closely by the desktop, computer keyboard and mouse. Surprisingly, the toilet seat was about 400 times cleaner than the desktop.

the laptop of luxury. They can feast all day from breakfast to lunch and even dinner.' Your desk is the second 'germiest' place in the office.

Pat Rusin from the University of Arizona is not sure why the shared toilet seat, which you would expect to be a maelstrom of maximum microbial activity, is actually one of the cleanest. He said: 'What we found, and what we are still theorising as to why, is that the toilet seat was always the cleanest site.' One theory is that toilet seats are too dry to provide a good home to a large population of bacteria.

The other major finding was that if you went to the trouble of using their sponsor's antibacterial wipes, you could drop the bacteria count by about 99.9%.

While I know intellectually that the toilet seat has a lower bacterial count than the desktop, I'm not going to have my next snack in the toilet. Perhaps I'll go halfway and wipe down my desktop, not with a germ-laden sponge (10 000 bacteria per square inch), but with a clean disposable tissue.

Flush Toilet with Lid Up or Down?

Dr Gerba also studied germ counts in the home and discovered the right way to flush the toilet. You should flush with the lid down.

If you flush with the lid up, a polluted plume of bacteria and water vapour erupts out of the flushing toilet bowl. He describes the germy ejecta as 'Baghdad at night during a US air attack'. The polluted water particles float for a few hours around your bathroom before they all land. Some of them will even land on your toothbrush.

Dr Gerba also found that in the home, the kitchen sponge had the highest germ count, followed by the kitchen sink. The lowest bacterial count, out of 15 household locations, was the toilet seat. He said (perhaps jokingly), 'If an alien came from space and studied the bacterial counts, he probably would conclude he should wash his hands in your toilet and crap in your sink.'

So if your toilet is in the bathroom and you flush with the lid up, you are probably brushing your teeth with toilet water. I guess that's one story to tell the males in your household, so that they put the lid down ...

References

Adams, Cecil, 'Does flushing the toilet cause dirty water to be spewed around the bathroom?', *The Straight Dope*, www.straightdope.com/classics/a990416.html.

Murphy, Cullen, 'Something in the water', *The Atlantic*, September 1997.

Woods, Kate, 'Toilet seats cleaner than desk', *Medical Observer*, 16 April 2004, p. 23.

Low Nicotine Cigarettes

About 90% of smokers know that smoking is bad for their health and about 60% of smokers want to give up, but can't. Every week, the average GP will hear an addicted smoker say, 'Look Doc, I'm already cutting down. I've started smoking low nicotine cigarettes.' These poor smokers have fallen for the myth that low nicotine cigarettes deliver less nicotine into their bodies.

Nicotine is an addictive drug. In fact, it's extremely addictive. Consider smokers who have had their larynx (voice box) removed as a result of a smoking-related disease. This is a very major operation. And yet, 40% of these smokers will, as soon as they recover from the operation, start smoking again.

Tobacco companies could make a zero-nicotine cigarette — but they won't, because nobody would buy them. After all, inhaling sugar does nothing for a cocaine addict. Addicted cigarette smokers need their nicotine hit.

A cigarette is an incredibly efficient drug-delivery device — luckily for the tobacco companies. It accurately delivers to the brain the precise dose of nicotine needed to ensure continued addiction within 11 seconds of sucking in the smoke. If you want to make somebody addicted, you need a very short time between when they do the action (sucking on the cigarette) and when they get the reward (nicotine in the brain). Smoking nicotine from a cigarette fits the bill perfectly. By the way, the average smoker carries about 40 billionths of a gram of nicotine in each millilitre of blood.

Nicotine can have opposite effects, depending on the dose. At low doses, it stimulates your thinking and increases your heart rate and blood pressure. At high doses, it calms you down and drops your heart rate. Smokers will often subconsciously adjust how hard and how frequently they suck, to get either a low or a high dose.

Will low nicotine cigarettes deliver low levels of nicotine? Yes, but only when they are tested on a sucking machine. However, when a human smokes low nicotine cigarettes, they get as much nicotine as they would from a regular nicotine cigarette.

Smokers need their regular nicotine hit, so when they change to a low nicotine cigarette, they just suck harder and more frequently — in order to get the same dose. Unfortunately, when you suck harder on a burning cigarette, you also suck in more carbon monoxide. This forces your body to make more haemoglobin, which makes your blood more 'sludgy' — which

I'm cutting back ... just after I finish this pack

Our friend and machine above are smoking low nicotine cigarettes ... Only one of them is addicted to nicotine and thus sucks harder to obtain the same 'hit' as full strength cigarettes.

Low Nicotine

Nicotine is an addictive drug. Low nicotine cigarettes do deliver less nicotine when 'sucked' on a controlled machine, however, the machine isn't addicted to nicotine ... and the human is. The result – humans smoke more furiously.

increases your chances of a stroke. So it's actually more dangerous to get your 'regular' nicotine dose from a low nicotine cigarette.

(Just as an aside, if a person who usually smokes medium nicotine cigarettes tries high nicotine cigarettes, they will suck less deeply, so that they still get their regular nicotine hit.)

The tobacco companies like low nicotine cigarettes. In 1978, the Imperial Tobacco Ltd wrote, in an internal document: '... the advent of ultra low-tar cigarettes has actually retained some potential quitters in the cigarette market by offering them a viable alternative ...'

So, low nicotine cigarettes not only deliver less nicotine, and give you more carbon monoxide, they also lull you into a false sense of security.

References

Bittoun, Renee, *The Management of Nicotine Addiction: A guide for counselling in smoking cessation*, University of Sydney Printing Service, 1998.

Fagerstrom, Karl Olov, 'Towards better diagnoses and more individual treatments of tobacco dependence', *British Journal of Addiction*, 1991, vol. 86, pp. 543–547.

Goldfish Memory

The Chinese had already domesticated the goldfish some thousand years ago (during the Sung Dynasty, 960–1279 AD). Since then, many centuries of selective breeding have given us over 125 types of goldfish. However, rumour has it that no goldfish can remember anything earlier than a few seconds ago. So every circuit of their tank or pond should be fresh and new — because they can't remember the last loop.

Do fish have a memory? And how can you determine whether a fish has a memory? Or whether any animal has for that matter?

Clearly, the Clark's Nutcracker has a superb memory. This bird lives in North America, and hoards food to get it through the winter. As autumn approaches, a single bird harvests up to 33 000 pine seeds. It then buries them in some 7000 separate hidden treasure troves, each with about four or five seeds. Its memory is so good that it is able to find each of these individual 7000 stockpiles later. It digs up and eats the seeds to survive the winter.

Few humans could do this — except perhaps Hiroyuki Goto, of Keio University in Tokyo, who in February 1995, recited *p* (the ratio of the circumference of a circle to its diameter) to 42 194 places.

Jonathon Lovell from Plymouth University's Institute of Marine Studies in the United Kingdom is convinced that some fish have a memory. He has successfully trained fish to swim towards a sound. He wants to release fish raised in captivity into the open sea, and call them back with special sounds to a feeding station, to supplement their natural diet.

Culum Brown (of the Institute of Cell, Animal and Population Biology at the University of Edinburgh) studied the Crimson-spotted Rainbow Fish while in Queensland. He compared fish that knew their tanks well with fish that had been just placed in tanks. He introduced a net with a central hole into a tank, and then swept it from one end to the other. The fish that had a strong memory of their tank were better able to escape through the central hole — presumably because they could ignore what they remembered as being familiar and non-threatening to them (their tank), and instead, could concentrate on the new threat (the net). The fish that knew their tank remembered the trawling net so well, that they could escape it in a follow-up study some 11 months later.

By the way, 11 months is nearly one-third of a goldfish's three-year life span — a very long time to remember something that has happened to you only once. In human terms, that's about 25 years.

Who the hell are you, and whatcha lookin' at?

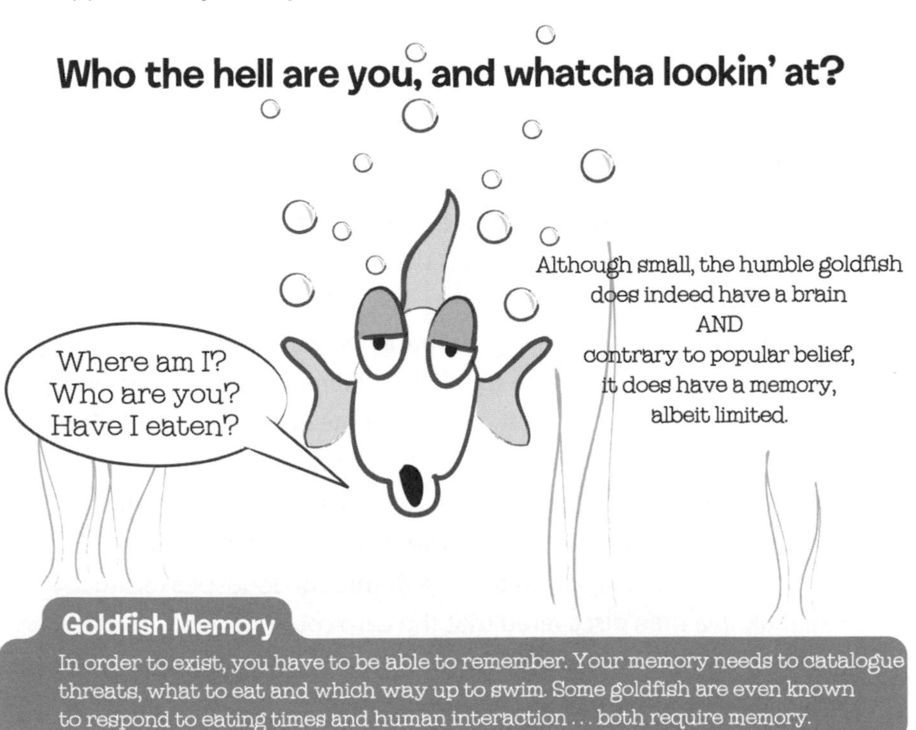

Goldfish Memory

In order to exist, you have to be able to remember. Your memory needs to catalogue threats, what to eat and which way up to swim. Some goldfish are even known to respond to eating times and human interaction ... both require memory.

Yoichi Oda of Osaka University in Japan has spent years studying the fine details of memory in goldfish — and he is also convinced that goldfish have a good memory.

Of course, there are thousands of anecdotes from owners of goldfish, who say that the fish remember regular feeding times. This is very impressive — after all, the goldfish food they are given looks nothing like the food that they are genetically programmed to eat.

Other owners say that their goldfish remember their faces and frolic about in the tank when their owners are the only ones present, but hide for an hour or so when strangers enter the room.

Different Types of Learning

Some goldfish will come to the glass of their tank whenever people walk into the room. These particular goldfish have worked out that when people turn up, so will food — at least, sometimes. In other words, people equal food. This is called 'associative learning'. The fish now associate people with food.

Some species of fish are very social and hang out together in schools. To survive in the school, they spend a lot of time paying attention to what their schoolmates do and learn by watching them. This is called 'social learning'.

Some fish can learn music — probably because, in the wild, it's important for them to be able to tell the difference between different sounds in their environment. Ava Chase of the Rowland Institute for Science in Cambridge, Massachusetts taught carp to tell the difference between John Lee Hooker's blues music and a classical oboe concerto by Bach, by feeding them smaller fish as a food reward. The music was played to the fish through loudspeakers in their tank. Ava then discovered that the carp could generalise from what they had learnt, and classify music that they hadn't heard before into the categories of blues or classical.

References

'In brief: musical fish', *New Scientist*, 19 January 2002, p. 24.

Brown, Culum, 'Familiarity with the test environment improves escape responses in the Crimson-spotted Rainbow Fish, *Melanotaenia duboulayi*', *Animal Cognition*, vol. 4, 2001, pp. 109–113.

Cellulite

Since the 1960s, the television, radio and print media has continually informed us that cellulite is the bane of the appearance-conscious person's life. And each time, the media usually feed us at least two of the four cellulite mythconceptions.

The first mythconception is that cellulite is abnormal and should therefore be removed. Second, cellulite is caused by toxins and/or poor circulation and/or clogged lymphatics. Third, if you get skinny enough the cellulite will simply disappear. And fourth, there's a revolutionary new product that will get rid of cellulite.

We all carry some fat as a percentage of our body weight — roughly 15–25% for men, and 20–33% for women. This fat is stored in fat cells. Millions of fat cells sit happily side by side, like a sea of soft butter balls. The fat in your body, like butter, doesn't have a lot of structural integrity, so you need a few fibrous bands running across this sea of soft bubbled fat to hold it together. Sometimes, there are so many fibrous bands crisscrossing the sea of fat balls that they turn this smooth sea into a set of many bumpy lakes. The technical term for the fibrous bands is 'hypodermal fibrous strands'.

Cellulite is lumpy-bumpy fat, stored in little pockets. *Stedman's Medical Dictionary* calls cellulite a 'colloquial term for deposits of fat and fibrous tissue causing dimpling of the overlying skin'. It is often found in the lower buttocks, and the backs of the outer thighs and hips.

Hail damage or natural texture?

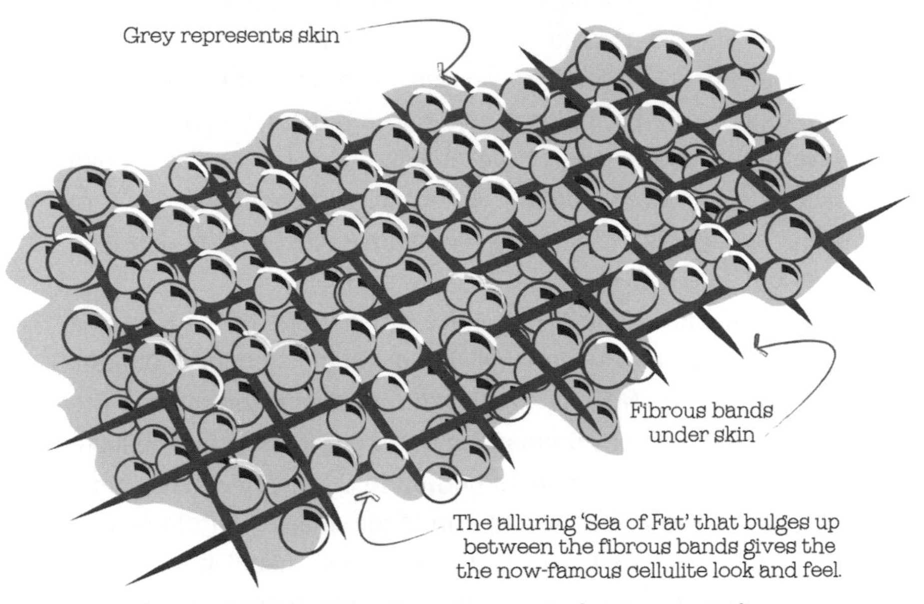

Normal: 75% of the female population has cellulite

Abnormal: only 25% of the female population don't have cellulite

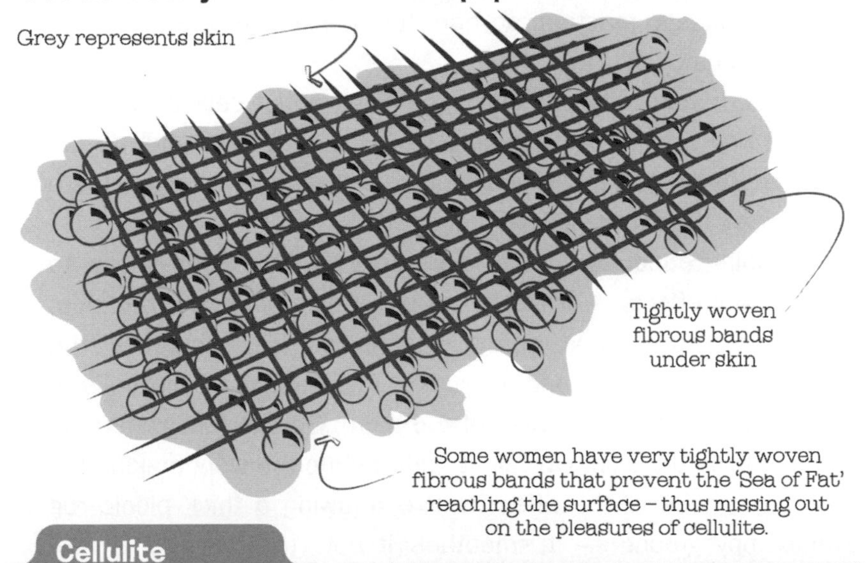

Cellulite

The mythconception is that cellulite is abnormal and should be removed. It is far from abnormal with 75% of women having it in one form or another. Statistics would therefore show it as being quite normal.

Cellulite is a normal way to store superficial fat. It has nothing to do with clogged lymphatic vessels, toxins or poor circulation. It is normal. The vast majority of women carry it. You can find cellulite in about 70–80% of women — even if they are really skinny. (Apparently, even ultra-svelte Nicole Kidman has a little.)

Given that most women have cellulite, this makes for a huge market — a merchant's dream. Anti-cellulite creams claim to turn dimpled, spongy skin (that looks like cottage cheese) into flat skin (as smooth as a baby's bottom).

These creams almost always have at least one of the following 'magic' ingredients — caffeine (a diuretic), tretinoin, dimethylaminoethanol (DMAE, an antioxidant) or aminophylline. The *European Journal of Dermatology* reviewed 32 anti-cellulite products. Some of the products had up to 31 ingredients — most of them fairly random apart from one of the 'magic four'. In fact, in the 32 products tested, 263 ingredients were used! It was as if the manufacturers were making wild guesses, and adding unproved products haphazardly. The other constant ingredient was some type of fragrance. About one-quarter of these products can cause an allergic reaction, so harmful side effects are a definite (but small) risk.

The dermatologists agree on one thing about cellulite creams — they won't make cellulite go away. On the other hand, if your skin is normally dry, and if your anti-cellulite cream incorporates a moisturiser, then your skin will look better.

Tretinoin (retinol) is another ingredient that has been shown to have helpful effects — in some people. It can penetrate the skin. It increases the blood supply of the dermis, increases the production of collagen, and increases the turnover of the cells in the epidermis (the outer layer of the skin). So tretinoin generates newer cells in the skin. Overall, it 'plumps' up the skin, making the outer layer thicker. The effect is like throwing a thick picnic rug over bumpy ground — it smoothes it out. The cellulite is now harder to see, because it's hidden under the 'plumper' superficial skin. On the other hand, all tretinoin products carry warnings about potential risks if used when pregnant.

The anti-cellulite effect is very small for tretinoin. And it's probably smaller for the remainder of the 'magic four' chemicals. (For example, the only 'research' that 'proves' that caffeine works was carried out by two companies making anti-cellulite products.) And the chemicals don't work on everybody.

At their very best, anti-cellulite creams produce a tiny improvement — perhaps.

There is one treatment that *does* work — liposuction, or removal of the fat cells. However, the remaining fat cells in the skin will grow larger to compensate — and you could be back where you started.

Professor Lisa M. Donofrio, a dermatologist at Yale University School of Medicine in the United States is sceptical of the creams. She says that we should just learn to live with, and love, our cellulite. Perhaps in another alternate dimension, she says, smooth skin is seen as boring while skin with lumpy-bumpy fat is seen as beautiful.

One Company's Opinion

A paper which discussed cellulite and what retinol (tretinoin) did to it was published by the *American Journal of Clinical Dermatology*. Two of the authors of the paper were from the Johnson and Johnson Consumer Report division.

They believed that cellulite is the result of two forces working against each other. One is a moderate, but long-term, excess deposition of fat. The opposing force is that the strands running across the balls of fat overreact to the presence of the fat thus becoming thicker and stiffer.

In the study, the 15 women who used retinol showed an average increase of about 10% in the elasticity of their skin. But 'the lumpy-bumpy appearance of the skin either showed little response, or was not responsive to the treatment'. The report also stated that: 'many purported cosmetic and medical treatments show little effect in improving cellulite, and certainly none cause its complete disappearance.'

References

Pierard-Franchimont, Claudine, *et. al.*, 'A randomized, placebo-controlled trial of topical retinol in the treatment of cellulite', *American Journal of Clinical Dermatology*, Nov–Dec 2000, pp. 369–374.

Sainio, Eva-Lisa, 'Ingredients and safety of cellulite creams', *European Journal of Dermatology*, December 2000, pp. 596–603.

Einstein Failed School

At the end of the 20th century, *Time* magazine voted Albert Einstein to be the Man of the Century. Albert was the guy who blew people's minds in the early 20th century with his Theory of Relativity (because Relativity was such a weird concept). Einstein was a genuine certified 'Mega-brain'. It is claimed that he even won the Nobel Prize for his work on Relativity.

It is also claimed that Einstein failed at school. This has consoled generations of school children with poor school marks. But both of these claims — Nobel Prize and failing at school — are dead wrong.

First, Einstein did not win the 1921 Nobel Prize in Physics for his work on Relativity. Part of the reason was that this theory, even in 1921, was still controversial.

Let's back up a little. In 1905, Einstein had the biggest year of his life. He wrote, with the help of his wife, Mileva, five ground-breaking papers that, according to the *Encyclopaedia Britannica*, 'forever changed Man's view of the Universe'. Any scientist would have been proud to write even one of these magnificent papers — but Albert published five of them in one year!

One paper, of course, dealt with Relativity — what happens to objects as they move relative to other objects. Two papers proved that atoms and molecules had to exist, based on the fact that you could see tiny particles jigging around when you looked at a drop of water through a microscope. A fourth paper looked at a strange

Sitzung vom 27 Juli 1900.

Anwesend die Herren Hurwitz, Weber, Fiedler, Minkowsky, Herzog, Franel, Geiser, Lacombe & Wolfer.

1). Das Protokoll der Sitzung vom 13 Juni wird genehmigt.

2). Der Vorstand gibt Kenntniss von zwei Mittheilungen des Schulrathes, wonach

a). Dem Studierenden Spiegler auf Antrag der Conferenz für die Lösung der Preisaufgabe ein Preis von 400 Fr. & die silberne Medaille ertheilt wird,

b). Die Studierenden Buff & Dompierre wegen Militärdienst, der erstere vom 21. Juli, der letztere vom 16. Juli bis zum Semesterschluss beurlaubt sind.

3). Quartalbericht. Buff erhält wegen Studienvernachlässigung einen Verweis durch den Vorstand. Stauber ist wegen Krankheit den grössten Theil des Semesters abwesend geblieben, hat infolge dessen keine Noten & wird den Curs wiederholen.

4). Schlussdiplomprüfungen. Die Ergebnisse der Prüfungen sind die nachstehenden:

	Funkt. Theorie	Geom.	Arithm. & Algebra	Theoret. Physik	Astron.	Diplom. Arbeit	Summe	Mittel
Ehrat	11	11	4 1/2	5	5	20	56 1/2	5.14
Grossmann	11	12	4	4 1/2	4	22	57 1/2	5.24
Kollros	12	11	4 1/2	4 1/2	6	22	60	5.45

	Theoret. Phys.	Prakt. Phys.	Funkt. Theorie	Astron.	Dipl. Arbeit	Summe	Mittel
Einstein	10	10	11	5	18	54	4.91
Marić	9	10	5	4	16	44	4.00

Mit Ausnahme von Fräul. Marić werden die sämmtlichen übrigen Candidaten zum Diplom empfohlen.

A. Wolfer.

The results of Einstein's and Mileva Maric's final exams. It's an extract from the minutes of the mathematical department (*Protokoll der Mathematischen Abteilung*). It is said that all candidates passed the exams successfully, except Mrs Maric.

property of light — the Photoelectric Effect. Plants and solar cells use the Photoelectric Effect, when they turn light into electricity. In fact, each year, plants (doing the Photoelectric Effect for free) turn 1000 billion tonnes of carbon dioxide into 700 billion tonnes of oxygen and organic matter. (And some people don't like plants!)

His fifth paper was a mathematical footnote to his Special Theory of Relativity. It was called 'Does the Inertia of a Body Depend on its Energy Content'. This paper carries the famous $E = mc^2$ equation, where E is the 'energy', m is the 'mass', and c is the speed of light. If you convert a mass, m, entirely into energy, this equation tells you how much energy you get. My children and I were lucky enough to see this equation, written in Einstein's hand, when a Special Relativity manuscript came to Sydney as part of a worldwide tour. I felt an amazing sense of awe.

The Theory of Relativity captured the public's consciousness. In the 1920s, there were claims that only five people in the whole world understood this theory. (Actually, these days a keen high school physics student could work through it.) But it was the unglamorous Photoelectric Effect that won Einstein the Nobel Prize. While he was in Shanghai he received a telegram from the Nobel Prize Committee informing him that he had been awarded the 1921 Nobel Prize for Physics 'for your photoelectric law and your work in the field of theoretical physics'. There was absolutely no mention of Relativity.

And now the second myth. Einstein definitely did *not* fail at high school.

Einstein, bring me your torch and leave the school immediately

Einstein – genuine certified 'mega-brain'

Einstein Failed

In 1896 – Einstein's last year at school in Aargau – the school's system of marking was reversed. A grading of '6', previously the bottom mark, was now the TOP mark. Einstein scored 4.91 out of 6 ... quite a good mark.

Einstein was born on 14 March 1879 in Ulm, Germany. The next year, his family moved to Munich, where he started school in 1886 at the age of seven. At the age of nine, he entered the Luitpold-Gymnasium. By the age of 12 he was studying calculus — an advanced subject normally studied by 15-year-old students. He was very good at the sciences. But, because the 19th-century German education system was very harsh and regimented, it didn't really develop his non-mathematical skills (such as history, languages, music and geography). In fact, it was his mother, not the school, who encouraged him to study the violin — and he did quite well.

In 1895, he sat the entrance examinations of the prestigious Federal Polytechnic School in Zurich, Switzerland. He was 16, two years younger than his fellow applicants. He did outstandingly well in physics and mathematics, but failed the non-science subjects, doing especially badly in French — he was not accepted. So he continued his studies at the canton school in Aargau, studied hard, and finally passed the entrance exams.

In October 1896, he finally began his studies at the Federal Polytechnic (even though, at 17, he was still one year younger than most of his fellow students). Also in that year, he wrote a brilliant essay that led directly to his later work in Relativity. Einstein did not fail at high school, and was definitely not a poor student.

So how did this myth start?

Easy. In 1896 — Einstein's last year at the school in Aargau — the school's system of marking was reversed.

A grading of '6', previously the bottom mark, was now the top mark. (Einstein scored 4.91 out of 6 — quite a good mark.) A grading of '1', previously the top mark, was now the bottom mark. Anybody looking up Einstein's grades would see that he had not scored any grades around '1' — which under the new marking scheme, meant a 'fail'.

School children can't use this mythconception as a crutch any more — they'll just have to work harder ...

Special Relativity for Idiots

Special Relativity is fairly easy to understand if you remember one thing — the only thing constant in the Universe is the speed of light. This is a slight oversimplification, but not too much.

Light travels at about 300 000 km/sec, or 300 m every microsecond (or millionth of a second).

Mass is not constant. As bodies travel faster, they get more massive. If they could reach the speed of light, their mass would be infinite (not just *big*, not just *as massive as the entire Universe*, but even bigger again — that is, *infinite*). Photons of light (which travel at the speed of light) get around this problem by having a mass of zero when they're not moving. When they are moving, they have a small mass.

Length is not constant. As a body travels faster, it shrinks (but only in the direction of travel) until it reaches zero length at the speed of light.

Time is not constant. As a body travels faster, its internal time slows down, until it reaches zero at the speed of light.

The only thing that stays constant in all of this is the speed of light.

Einstein's Brain

Einstein's brain went 'missing' soon after he died in 1955. Thomas Harvey, a duty pathologist at Princeton Hospital, New Jersey, removed the brain within seven hours of Einstein's death and preserved it. It then became the subject of controversy, because the executor of Einstein's will, Otto Nathan, claimed that Harvey was a thief.

Harvey left Princeton, and the brain 'vanished' until 1978, when the journalist Steven Levy found Harvey in Wichita, Kansas — and

Einstein's brain in a box marked 'Costa Cider'. Harvey didn't really have the qualifications to study Einstein's brain, so he had begun to mail small sections of it to expert neuroscientists.

A brain has 'neurons' (so-called thinking cells) and 'glial' cells (which supposedly don't do any thinking, but just act as 'support' cells to the neurons). Einstein's brain looked average to the naked eye, and was of average weight. Under the microscope, it had a very high ratio of glial cells to neurons in the inferior parietal lobe — part of the brain that does spatial and mathematical reasoning. To the trained eye, the inferior parietal lobe was about 15% bigger than normal.

Einstein's brain is finally back in Princeton — in a secret hiding place.

References

Reader's Digest Book Of Facts, Reader's Digest Pty Ltd, 1994, pp. 234, 416–417.

Broks, Paul, 'The adventures of Einstein's brain', *The Australian Financial Review*, 28 March–1 April 2002, p. 3.

Weiss, Peter, 'Getting warped', *Science News*, vol. 162, 21 and 28 December 2002, pp. 394–396.

All White, My Sun

Practically every society on our planet shares the mythconception that the Sun is yellow. These societies include Native Americans, Australian Aborigines (what colour is the Sun on the Aboriginal flag?), the Dutch and so on. Of course, they have all been deceived.

A white surface takes on the colour of the local lighting, and you can use this to prove that the Sun is white. Think about what happens if you wear white clothes in a nightclub with red mood lighting. Your clothes look red. What happens to a white car around midday on a sunny day? The white car remains resolutely white, refusing to turn yellow. The Sun is white. In fact, the Sun defines the word 'white'.

How did this almost-universal myth originate?

We don't really know, but one explanation is that the Sun is yellowish only when you can almost safely look at it — when it's on the horizon, at sunrise or sunset. At these times the light from the Sun has to pass through a much greater than normal thickness of air. The dust in the air bends away the blue light, leaving the other end of the spectrum — the yellow-red colours.

Or perhaps the Sun 'appears' yellow when you compare it to the blue sky. You get this effect if you stare at the blue sky for a long while, and then quickly take a glimpse of the Sun. Your colour vision has been shifted to the blue, and for a few moments, you see the Sun, and its afterimage, as yellow-red.

Regardless of how this myth started, it is taught to young children when they learn to paint. They are never told to 'paint the Sun white', but to 'paint the Sun yellow'. There is probably a very practical reason for this — white paint doesn't show up very well on white paper.

Consequently nearly all of us believe that the Sun is yellow, even though we see white cars and white clothes every day. They wouldn't look white if the Sun was actually yellow. And the Sun always looks white when you get a brief glimpse of it (as long as it's high enough above the horizon).

Why are so many of us tricked? We are a little like the characters in *The Matrix* movies, whose entire sensory systems are fooled! The only ones who have seen the light are the astronomers.

All White, My Sun

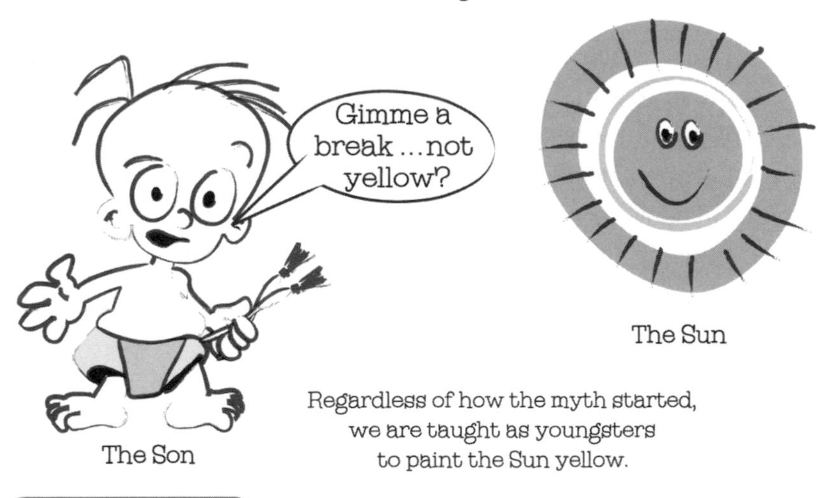

Regardless of how the myth started, we are taught as youngsters to paint the Sun yellow.

White Sun

Practically every society on our planet shares the mythconception that the Sun is yellow. These societies include Native Americans, Australian Aborigines, the Dutch and so on. Of course, they have all been deceived.

White Sun

The Sun pumps out its energy virtually across the whole electromagnetic spectrum. This energy includes x-rays, radio waves, and roughly in the middle of the spectrum, visible light. Sunlight appears to be white, because it consists of roughly equal intensities of all visible wavelengths, from red to blue.

Sir Isaac Newton was one of the first scientists to prove this. In 1665 and 1666, he experimented with sunlight coming into a darkened room, through a single small entrance. He placed a triangular glass prism in the path of the beam. When the light passed through the prism it split into all the colours of the rainbow, and landed on a white card. He had proved that white light is made up of the colours of the rainbow.

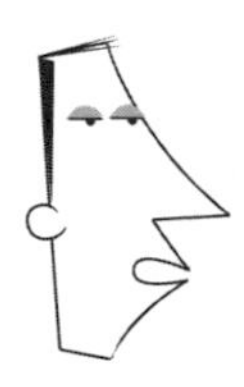

References

de Grasse Tyson, Neil, 'Things people say: The only thing worse than a blind believer is a seeing denier', *Natural History*, July–August 1998.

Fan Cools Room

One of the great summer myths is that a fan will cool a room. A fan cannot cool a room — it can only cool the people who are sitting in the room.

Think about a day when the air is cooler than your skin. On average, your body generates about 100 watts of waste heat (about as much as a light bulb). If there's no wind, this heat creates a thin layer of warm air that sits next to your skin. Once this layer has warmed up to skin temperature, it becomes a very good heat barrier. Heat cannot pass into this layer from your skin — because heat can normally travel only from a hot place to a cooler place. As your skin temperature increases, you begin to feel uncomfortable. The only way out of this cycle is to move around, or to sweat.

Enter the fan.

When a fan blows wind across your skin, it pushes away this warm layer of air. On a day when the air is cooler than your skin, the fan replaces the warm skin layer of air with cooler room air that has not been preheated by your skin. You will definitely feel more comfortable when the fan switches on. However, the fan did not cool the room. All it did was remove the warm air near your skin. You can prove that the fan didn't lower the temperature by checking a thermometer both before and after you switch the fan on. You will see that there is no change.

But what happens on a really hot summer's day, when the air is already hotter than your skin temperature? In this case, you begin to sweat. The fan will now cool you down even better. The air will blow over your slightly moist skin, and turn the water in your sweat into vapour, which it carries away. It takes a lot of energy to turn water-as-a-liquid into water-as-a-vapour. Because this energy comes from your skin, your skin will feel a lot cooler.

You can test this by wetting the tip of your finger in your mouth, and then blowing on it vigorously — it will feel cooler as the air carries away the molecules of water. Once again, you can prove that the fan doesn't cool the room at all by using a thermometer as you switch on the fan — the temperature doesn't change a bit.

By the way, evaporation is how a resting person dumps 25% of their waste heat. Water is evaporated from the lining of their lungs, as they breathe out.

This tells you two things. First, if you have a pet that does not perspire (such as a ferret), there's no point in using a fan to cool it down. (If your ferret is at risk of heat stroke, gently bathe it with tepid water.) Second, there's no point in leaving the fan running

Waitin' for the coolin' breeze

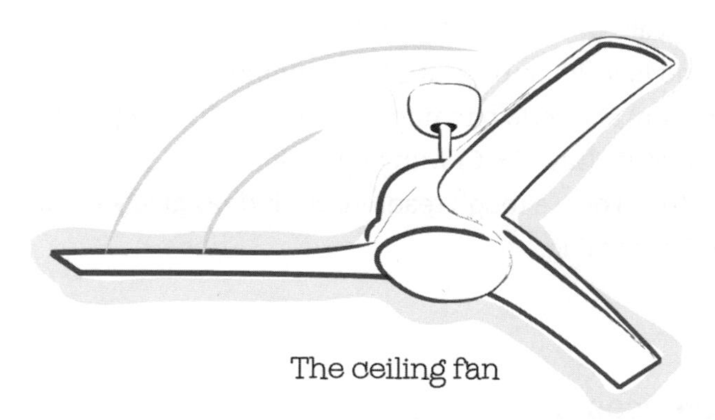

The ceiling fan

Cooling Fans

One of the great summer myths is that a fan will cool a room. A fan cannot cool a room . . . it can only cool the people who are sitting in the room by evaporating their perspiration.

when you're not in the room. In fact, the motor in the fan will generate a little waste heat, making the room slightly warmer.

Extreme Air Cooling

A fan cools you by removing a layer of warm air that would normally sit next to your skin. This is fine in the middle of summer, but could kill you in the middle of winter.

The technical term for this is 'wind-chill'. Antarctic and Himalayan travellers have found that they can survive perfectly well, bare-chested, in temperatures as low as –40°C — as long as there is absolutely no wind blowing. But if there is the slightest breath of wind, they have to cover up immediately.

The wind removes the warm air and replaces it with cold, dry air. It takes a lot of energy to warm and moisten this new air — more than you can deliver easily.

The wind-chill factor was created in the mid-1940s to give people living in cold climates an easy way to work out what effect a wind would have. A scale was devised to predict the combined effect of wind speed and low air temperature. For example, a wind of a certain speed (say 16 kph) at a certain temperature (say –4°C) is equivalent to a lower temperature (say –13°C) with no wind.

Of course, this is only a rough reading, but it does give you some kind of effective warning.

References

Holper, Paul N., *Wow! Amazing Science Facts and Trivia*, ABC Books, Sydney, Australia, p. 75.

Vondeling, John, *Physics, A World View*, Saunders College Publishing, USA, p. 222.

Walker, Jearl, *The Flying Circus of Physics*, John Wiley & Sons Inc., USA, p. 51.

Eclipse Blindness

A solar eclipse occurs when the Moon passes between the Earth and the Sun. In a total eclipse, the Moon blocks out all of the sunlight, creating an eerie deep twilight. Suddenly, in the middle of the day, you can see the stars. But many people, when given the chance, never enjoy the free cosmic thrill of a total solar eclipse, because they believe the myth that looking at a solar eclipse, or even being outdoors when it happens, will make you go blind. In fact, a total eclipse of the Sun can be pretty harmless.

A *partial* eclipse, however, is far more *dangerous*. In a partial eclipse, the Moon blocks only some of the sunlight. (Actually, if you were not aware of the eclipse, you would probably think that a cloud was just temporarily covering the Sun.) But even if 99% of the Sun is covered, the tiny crescent of Sun remaining is bright enough to blind you, if you stare at it for anything longer than the briefest moment.

The Sun does not emit new and strange forms of damaging radiation during an eclipse but continues to squirt out what it always has. You can certainly damage your eyes by staring at the Sun when it is partially covered by the Moon, because it is still emitting enough energy to damage you. You have to remove about 99.9968% of the Sun's energy to make it safe to look at.

The Sun gives out heat energy as well as light. The heat energy is focused and concentrated onto the central part of your retina, which deals with fine vision. If you stare at the Sun for long

I've been eclipsed

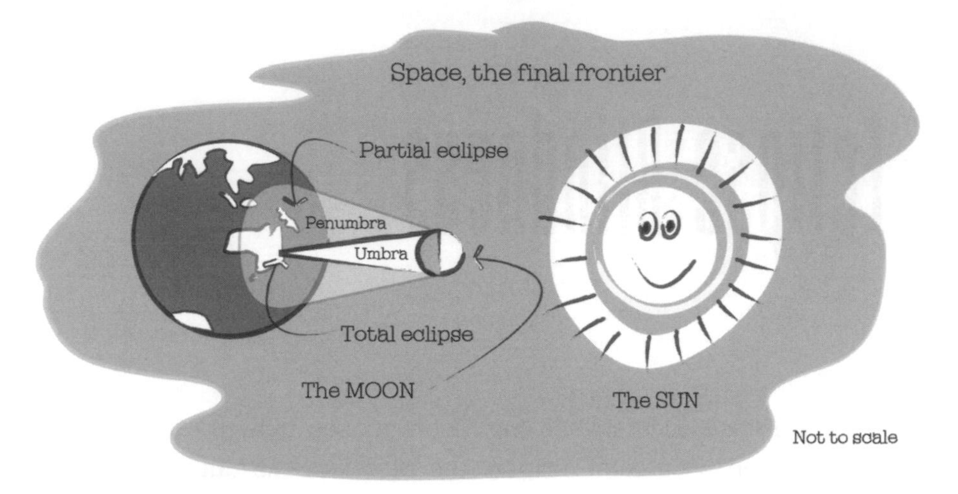

A solar eclipse occurs when the Moon passes between the Earth and the Sun. In a total eclipse, the Moon blocks out all of the sunlight to an area, creating an eerie deep twilight.

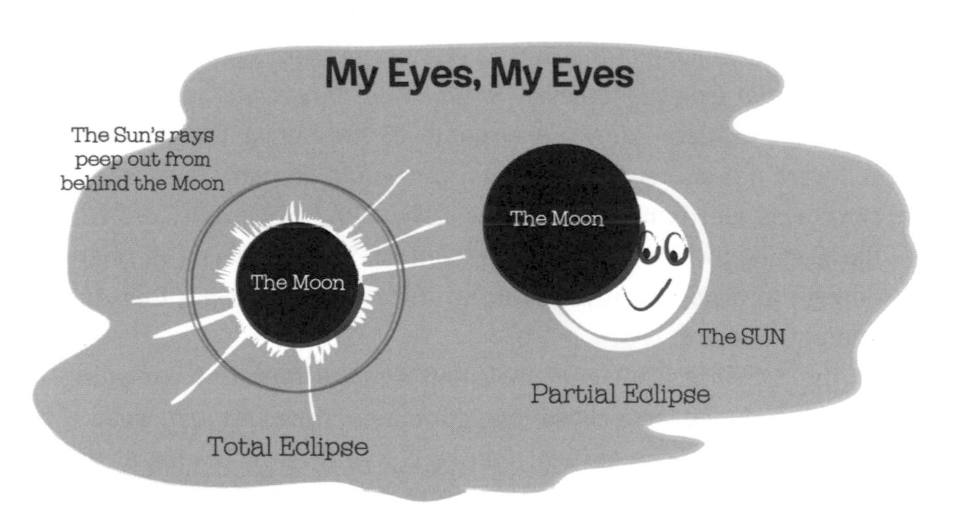

Eclipse Blindness

If you look at the Sun with the naked eye, you will burn out the centre of the retina. If you look at the Sun through a brick wall, no light will reach your retina and you are safe. Somewhere between 'no protection' and 'the brick wall', is a safe level.

enough, you will lose your central vision. The Sun's energy will burn out the central part of the retina. In extreme cases, the tissue will literally cook. You usually won't notice this happening, because there are no pain receptors in the eye. Because your peripheral vision will not be affected, you will still be able to see out of the corner of your eye. But if you try to read fine print, it will be fuzzy, like reading through a metre of seawater, or a glass smeared with petroleum jelly.

The US Army documented a partial solar eclipse causing blindness on 4 February 1962, with military personnel in Hawaii. The partial eclipse had enough energy in the exposed sliver of Sun to burn out the centres of the soldiers' retinas, when they looked at the Sun. Over the next few days, many of the soldiers had trouble shooting accurately on the rifle range. Their vision had dropped from 20/20 to 20/200 — 10 times worse than normal. Most recovered, but some had permanent loss of visual sharpness.

When the next total eclipse of the Sun rolls around, just remember a few rules:

- Observing the totally-eclipsed Sun with the naked eye is fine — but only when the Sun is totally covered by the Moon. You have to pick the moment, when there is no direct solar radiation to damage the eye.
- Never look at the partially-eclipsed Sun with the naked eye. Even a slim crescent of Sun has enough energy to blind you.
- It is safe to look at the fully-exposed or partially-exposed Sun using approved filters, such as professional Solar Viewing Mylar filters. Never look directly at the Sun through smoked glass, exposed photographic film, magnetic discs ripped out of old floppy discs, CDs, or Mylar food packaging.

The pinhole camera method is another safe way to view an eclipse in action. Punch a two-millimetre hole into a piece of card and, with your back to the Sun, hold the card so that the Sun's light passes through the hole and onto another card which acts as a screen. The image of the Sun will be slightly fuzzy but you will be able to see the shape of the eclipsed Sun.

As long as you don't look directly at the Sun, it is perfectly safe to be outdoors as the Sun drifts in or out of the eclipse.

It is also perfectly safe to look at the totally-eclipsed Sun. This is because the Moon, which is 400 times smaller than the Sun, is also 400 times closer — so it can totally obscure all direct light from the bright part of the Sun. In this brief window of totality, it is safe to abandon the pinhole camera briefly, and gaze up in awe at the total eclipse of the Sun, complete with the shimmering corona, and all its associated splendours.

How Dark is Safe?

If you look at the Sun with the naked eye, you will burn out the centre of the retina. If you look at it through a brick wall, no light will reach your retina and this will be perfectly safe.

Somewhere between 'no protection' and the 'brick wall', is a safe level. For the case of the visible and infrared light coming from the Sun, the safe level is provided by a filter (Shade 12) that lets through 0.0032% of the light. But this is still a little bright for most people, so a darker filter (Shade 14) that transmits 0.0003% of light is more comfortable.

References

Harrington, Philip S., *Eclipse!: The what, where, when, why and how guide to watching solar and lunar eclipses*, John Wiley & Sons Inc., USA, 1997, pp. 2–3, 129–131, 205–208.

Chou, Ralph, 'Solar filter safety', *Sky & Telescope*, February 1998, pp. 36–40.

Anaesthetic Bomb

The 'make-them-unconscious' Anaesthetic Bomb has been appearing in movies for over half a century. Because movie 'good guys' never kill anybody, they pull out the Anaesthetic Bomb, and roll or slide it towards people, who immediately fall to the floor unconscious. The Anaesthetic Bomb was used in the 2002 remake of *Ocean's Eleven* — and you can be sure that it will appear in future movies.

The Anaesthetic Bomb has (according to legend) been used in house robberies by the villains who can then make their way through a house of sleeping residents without waking them. And of course, its use has been proposed in commercial jetliners as a way to harmlessly subdue violent passengers or terrorists.

The Anaesthetic Bomb is not designed to affect just a region (e.g. regional anaesthesia for a leg) or a small patch (e.g. local anaesthesia for torn skin). No, in the movies, the bomb needs to give general anaesthesia instantaneously, making the person unconscious instead. A person under general anaesthesia is unaware of their environment, cannot feel pain, cannot move and has no memory of what happens during the anaesthesia.

General anaesthesia took a long time to invent. In the *Odyssey*, Homer wrote about an Egyptian herbal drink called *napenthe* (probably either opium or cannabis) which eased grief and banished sorrow. In 1799, the English chemist Sir Humphry Davy, discovered that laughing gas (nitrous oxide) could relieve the pain of his infected tooth — but his claim was ignored. By 1842, an American surgeon

Crawford Long, had started using ether as an anaesthetic, but he did not publish his results until 1849. Therefore the first use of true surgical anaesthesia is usually attributed to William Morton, an American dentist who, in October 1846, administered ether to a patient having a neck tumour removed at the Massachusetts General Hospital in Boston.

In the early days of anaesthesia, usually only one drug was given. But today, several drugs are given at the same time to achieve many different aims (such as unconsciousness, muscle relaxation, loss of memory, paralysis and immobility). However, because these drugs travel via the blood to the brain, they can affect other organs in the body along the way. For this reason the anaesthetist has to monitor very closely the patient's heart rate and rhythm, blood pressure, breathing rate and oxygen level in the blood.

There is a very fine line between being conscious, being in a state of general anaesthesia, and being dead.

The 'Make-Them-Unconscious' Bomb (as seen on TV)

Movie 'Good Guys' never like to be seen to kill anybody. The answer is the 'near' harmless, Acme-patented Anaesthetic Bomb. Simply pull the pin, roll or slide it towards baddies or villains and they will immediately fall to the floor unconscious.

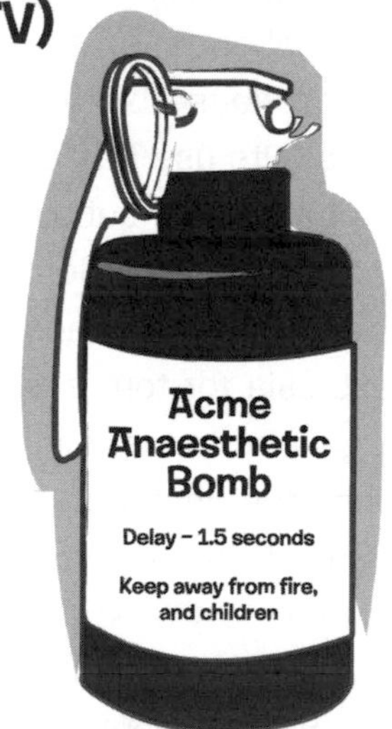

Anaesthetic Bomb

For the 'bomb' to work in the movies, it needs to deliver a general anaesthetic instantaneously, making the person / people unconscious. This has yet to be developed for the real world.

It is very difficult to render someone unconscious without injuring or killing them. Doctors have to study medicine for five years or more, and then spend several years in general hospital training. To become an anaesthetist, a further four years of study is required.

If safe and effective Anaesthetic Bombs really did exist, life would be easy for anaesthetists. They wouldn't have to spend over a decade studying — all they would need to know is to be out of the room when the bomb goes off.

Moscow Anaesthetic Bomb

On Wednesday, 23 October 2002, over 800 people were watching *Nord Ost*, a classic Russian musical in the Moscow Musical Theatre. Suddenly, over 50 Chechen militants, armed with machine guns, rushed in, held the audience hostage, and laid explosives inside the theatre and at all the exits. They then demanded that Russia withdraw its army from Chechnya, or else they would kill the hostages.

Just before dawn on Saturday, 26 October, Olga, a 21-year-old survivor, saw billowing clouds of a grey gas drift into the theatre. She covered her face with her scarf and dropped to the floor — and was one of the few hostages who did not lose consciousness. Almost immediately, Russian soldiers stormed the theatre, shooting the Chechens.

Even two days after the rescue, two-thirds of the surviving hostages were in hospital in a serious condition. About 117 hostages died from the effects of the still-unknown anaesthetic gas.

References

Schiermeier, Quirin, 'Hostage deaths put gas weapons in spotlight', *Nature*, vol. 420, 7 November 2002, p. 7.

Rieder, Josef, *et. al.*, 'Moscow theatre siege and anaesthetic drugs', *The Lancet*, vol. 361, 29 March 2003, p. 1131.

Everest Not Tallest Mountain

Most Australian school students are taught that the highest point on Earth is the tip of Mt Everest and that Mt Kosciusko is the highest mountain in Australia. But most school students have been misled.

So what is the story on Mt Kosciusko? It is indeed the highest mountain (at 2228 m) on the Australian *mainland*.

But the highest recognised Australian mountains are in the Australian Antarctic Territory — Mt McClintock (eastern sector — 3490 m) and Mt Menzies (western sector — 3355 m). However, the highest mountain on Australian Sovereign Territory is Mt Mawson (2745 m) in the Big Ben mountain complex on Heard Island in the Southern Indian Ocean, about 4000 km southwest of Perth. So while Kosciusko does have the distinction of being the highest mountain on the Australian mainland, it isn't the highest mountain on Australian territory.

And we do have volcanoes.

Australia doesn't have the technology to monitor the volcanic activity of weather-bound and remote Heard Island, but there were some very spectacular eruptions on Big Ben in February 2001.

What about Mt Everest? Is it the highest mountain in the world? It all depends on what you mean by 'highest'. Does it mean the 'highest above sea level', or does it mean that it 'pokes out most into space and is furthest from the centre of the Earth'?

Back in the 17th and 18th centuries, it was thought that a certain Mt Chimborazo, an extinct snow-capped volcano in Ecuador, was the highest point on Earth, at 6310 m above sea level. In 1852, the Great Trigonometrical Survey of India noted that a certain mountain named Peak XV was the highest at 8840 m. The British named it Everest in 1865, after Sir George Everest, who was the British Surveyor General from 1830–1843. It did not seem to matter that the local Tibetans and Nepalese had already given the mountain some perfectly good names — *Chomolungma* or 'Mother Goddess of the Land' by the Tibetans, and *Sagarmatha* by the Nepalese. Indeed, Everest himself thought that the mountain should keep its local name — but he obviously didn't protest too loudly.

The height of Mt Everest was adjusted to 8848 m in 1955, and then to 8850 m in 1999, after a team of climbers used state-of-

The High Point of Any Discussion

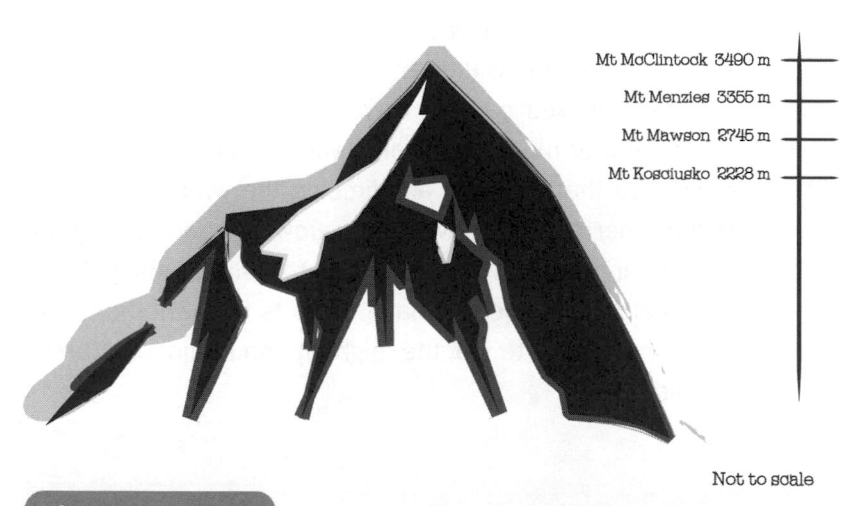

Highest Peaks

While Mt Kosciusko has the distinction of being the highest mountain on the Australian mainland, it isn't the highest mountain in the Australian territories.

the-art satellite measuring devices on the summit. All of these heights are measured above sea level.

The reason that Mt Everest is not the highest point on Earth is that the Earth spins — and this spin makes the whole planet bulge outwards at the equator. The diameter of the Earth through the equator is about 21 km more than the diameter of the Earth measured through the North and South Poles.

Let's look again at Mt Chimborazo, which was once thought to be the highest mountain on Earth. It was first climbed by Edward Whymper in 1880.

Mt Chimborazo is about 1.5° south of the equator, while Mt Everest is much further around the curve of the Earth at 28° north. So although Mt Chimborazo is about 2540 m closer to sea level than Mt Everest, it is about 2202 m further away from the centre of the Earth. It pokes further into space than Mt Everest. If this were better known, perhaps the achievements of the Conqueror of Chimborazo, Edward Whymper, would have made more of a bang. (In fact, three other peaks — Huascaran, Cotopaxi and Kilimanjaro — are also 'higher' than Mt Everest.)

However, Mt Everest is still the highest mountain above sea level. By the end of 2001, some 1314 people had reached its summit, and 167 people had died trying. If you have US$65 000, and are physically fit, you can try to reach the peak. But if you don't have that kind of money to spend, you can console yourself with the knowledge that they are all climbing the wrong mountain anyway. On the other hand, Mt Everest is growing about 5–10 mm each year as the Indian land mass rams into Asia, pushing Tibet higher. So all that wealthy people have to do is wait another half-million years for Everest to be the highest mountain on Earth, under all definitions ...

High History

After Mt Chimborazo, the title of 'highest mountain in the world' was handed to the Himalayan peak Dhaulagiri (8172 m) in 1809, and then to Mt Kanchenjunga (8598 m) in 1840.

Mt Everest was first climbed on 29 May 1953, by Edmund Hillary of New Zealand and the local Sherpa mountaineer Tenzing Norgay. Neither has ever admitted who was technically the first to reach the peak.

Heard Island (which has the highest mountain on Australian Sovereign Territory) was discovered by a British sealing vessel in 1833. It was later named for Captain John J. Heard, an American mariner. In 1947, control of Heard Island was transferred from the British to the Australian government.

References

'2 of British team conquer Everest — highest peak won', *New York Times*, 2 June 1953, p.1.

'Ask us', *National Geographic*, January 2002.

Kiernan, Kevin, Fitch, Stu & McConnell, Anne, 'Big Ben: the fire beneath the ice', *Australian Antarctic Magazine*, Spring 2001, pp. 4–5.

Pott, Auriol, 'A friend told me that Mt Everest isn't the highest point on Earth. Is she right?', *Focus*, December 2000, p. 34.

Lemmings Suicide

One myth deeply entrenched in our language is that of the 'lemming suicide plunge' — where lemmings, apparently overcome by deep-rooted impulses, deliberately run over a cliff in their millions, to be dashed to their deaths on the rocks below, or to drown in the raging ocean. Indeed, this myth is now a metaphor for the behaviour of crowds of people who foolishly follow each other, lemming-like, to their inevitable doom. This particular myth began with a Disney movie.

Lemmings are rodents. Rodents have been around for about 57 million years and today make up about half of all the individual mammals on Earth. There are four genera of lemmings — Collared Lemmings, True (or Norway or Norwegian) Lemmings, Wood or Red-Backed Lemmings and Bog Lemmings. They are found in the cooler northern parts of Eurasia and North America. The True Lemming (which has the most impressive migrations) is about 10 cm long, with short legs and a short tail.

Many rodent species have strange population explosions. One such event occurred in the Central Valley of California in 1926–27 with the mouse population reaching around 200 000 per hectare (about 20 mice per square metre). In France between 1790 and 1935, there were at least 20 mouse plagues. But lemmings have the most regular fluctuations — these population explosions happen every three or four years. The numbers rocket up, and then drop almost to extinction. Even after three-quarters of a century of intensive research, we do not fully understand why their populations fluctuate so much. Various factors — change in food availability, climate, density of predators, stress of overcrowding, infectious diseases, snow conditions and sunspots — have been proposed, but none completely explain what is going on.

In the 1530s, the geographer Zeigler of Strasbourg, tried to explain these variations in populations by saying that lemmings fell out of the sky in stormy weather, and then suffered mass extinctions with the sprouting of the grasses in spring. In the 19th century, the naturalist Edward Nelson wrote that 'the Norton Sound Eskimo have an odd superstition that the White Lemming lives in the land beyond the stars and that it sometimes comes down to the earth, descending in a spiral course during snowstorms. I have known old men to insist that they had seen them coming down. Mr Murdoch records this same belief as existing among the Point Barrow Eskimo.' But none of the Inuit (Canadian Eskimo) stories mention the 'suicide leaps off cliffs'.

Follow the leader...

Lemmings Suicide

This particular myth began with a Disney movie, 'White Wilderness', which was released in 1958. A migration scene was filmed and the cliff-death-plunge sequence was done by herding the lemmings over a small cliff and into a river.

When these population explosions happen, the lemmings migrate away from the denser centres. The migrations begin slowly and erratically, evolving from small numbers moving at night, to larger groups in the daytime. The most dramatic movements happen with the True Lemmings. Even so, they do not form a continuous mass, but instead travel in groups with gaps of 10 minutes or more between them. They tend to follow roads and paths. Lemmings avoid water, and will usually scout around for a land crossing. But if they have to, they will swim. Their swimming ability is such that they can cross a 200-m body of water on a calm night, but most will drown on a windy night.

So there is a tiny kernel of truth in this myth. Lemmings have regular and wild fluctuations in population numbers — and when the numbers are high, the lemmings do migrate.

The myth of mass lemming suicide began when the Walt Disney nature documentary, *White Wilderness* was released in 1958. It was filmed in Alberta, Canada, far from the sea in a region not native to lemmings. So the film makers imported Collared Lemmings, by buying them from Inuit children. They filmed the migration sequence by placing the lemmings on a spinning turntable covered with snow, and then shooting it from many different angles. The cliff-death-plunge sequence was done by herding the lemmings over a small cliff into a river. It's easy to understand why the film makers did this. For one thing, wild animals are notoriously uncooperative. For another, a 'migration-of-doom' followed by a 'cliff-of-death' sequence is far more dramatic to show than the lemmings' self-implemented population-density management plan.

There are two mistakes here. First, it is the True Lemming that does the spectacular migrations, not the Collared Lemming shown in the documentary. Second, lemmings do not commit mass suicide. Instead, animals live to thrive and survive.

It's funny that Disney could use a rodent called Mickey Mouse as their mascot (who in the early days, generated much of their income), but still be so unkind to another rodent, the lemming …

Boom and Bust

Up where the lemmings live, the locals have a saying, 'Lemmings cycle, unless they don't'. This saying is very true, and also symbolises that we still don't understand why their population numbers go through boom-and-bust cycles. In a big year, lemming numbers can increase by a factor of 1000.

In 1924, the eminent British ecologist, Charles Elton, published an early paper describing how rodent populations varied wildly from year to year. Since then, ecologists have tried to understand why this occurs — with not a lot of success.

In October 2003, Olivier Glig (from the University of Helsinki) and his colleagues published their research. Over a 14-year window they looked at lemmings and their predators in the Karup Valley of Greenland. The population numbers followed a four-year cycle.

Three of the predators — Arctic Foxes, Snowy Owls and the Long-tailed Skua — migrated in and out of the area depending on the food supply, while the fourth predator — the stoat — stayed in the area. Glig's mathematical model predicted that the population numbers of the migratory predators would exactly follow the population numbers of the lemmings — and this is what they found. They also predicted, and found, that the population peak of the stoats came one year after the population peak of the lemmings.

We are getting closer to understanding boom-and-bust population numbers in some animals.

References

Brook, Stephen, 'Lemming myths takes fall', *Weekend Australian*, 1–2 November 2003, p. 15.

Gilig, Olivier, Hanski, Ikka & Sittler, Benoît, 'Cyclic dynamics in a simple vertebrate predator–prey community', *Science*, vol. 302, 31 October 2003, pp. 866–868.

Hudson, Peter J. & Bjørnstad, Ottar N., 'Vole stranglers and lemming cycles', *Science*, vol. 302, 31 October 2003, pp. 797–798.

Moffat, Michael, 'Do animals commit suicide?', *Discover*, July 2002, p. 12.

Stitch

If you have ever done any strenuous exercise and pushed yourself really hard, you'll probably have felt a 'stitch' — usually in the tummy. The Lack of Blood Theory of the stitch is fairly straightforward. It claims that because the exercise diverts blood to your arms and legs, there is less blood available for your central organs, especially the diaphragm muscle, which pulls your lungs downward so that you can breathe. The lack of blood causes the pain in your tummy, in much the same way that lack of blood to the heart muscle causes the pain of angina. Interesting theory — but it appears to be totally wrong.

The experts in this area — the exercise physiologists — don't call this pain a stitch. They call it ETAP, or 'exercise-related transient abdominal pain'.

The pain of a stitch usually happens in the abdomen, and most commonly, in the right upper quadrant near the liver. However, you can also get what athletes call a 'shoulder stitch'. In fact, I consistently get it at the 10-km mark in the 14-km Sydney City-to-Surf Fun Run. People assume that once you become fitter, you stop getting stitches — but stitches happen to one-fifth of the extremely fit runners in the 67-km Swiss Alpine Ultra Marathon. About 80% of stitches are a sharp pain, while 20% are a dull pain. Usually the pain subsides a few minutes after you stop exercising, but occasionally it can last for two or three days.

One problem with the Lack of Blood to the Diaphragm Muscle Theory is that you can get a stitch while doing activities that do not involve a lot of laboured breathing, such as motorbike riding or the ever-popular pastime of camel riding. And in fact, during the practice for the invasion of France in World War II, soldiers who were subjected to rough jolting while standing still in torpedo boats suffered stitches.

This fits with the second theory for the cause of stitches — Mechanical Stress on the Visceral Ligaments.

The organs in your gut (between the bottom of your ribs and the top of your legs) are held in place by many different ligaments. This theory says that these ligaments get strained by the continual up-and-down pounding of the weight of the internal organs. This fits in with why you supposedly are more likely to have a stitch while exercising after eating a full meal. However, it doesn't explain why one-fifth of expert swimmers still suffer from

You're like a pain in my abdomen

The 'stitch' suffering athlete

The Stitch

If you've ever done any strenuous exercise and pushed yourself really hard, you'll most likely have experienced the feeling of a 'stitch'... usually in the tummy.

stitches. After all, there's very little pounding as they power through the water.

But Dr Darren Morten, Director of the Avondale Centre for Exercise Sciences in New South Wales, has a third theory — Irritation of the Parietal Peritoneum.

The parietal peritoneum is a membrane that lines the entire gut cavity, including the bottom of the diaphragm muscle. Twisting your torso while swimming can irritate this membrane. A full stomach can also irritate your parietal peritoneum — in two separate ways. First, it places extra physical pressure on this membrane. Second, a full stomach sucks water into itself to aid digestion, dehydrating the parietal peritoneum. The Parietal Peritoneum Theory also explains the shoulder-tip stitch. Both the diaphragm near the liver, and the tip of your right shoulder, send their 'pain' signals to the same part of your brain. So you 'feel' some diaphragm pain as shoulder-tip pain.

So what can you do about a stitch? First, avoid doing strenuous exercise for at least two hours after a heavy meal. Second, avoid heavily sweetened drinks, and instead, drink isotonic fluids that have 6% carbohydrate.

References

'Last Word', *New Scientist*, 18 October 1999, p. 57.

Villazon, Luis, 'What causes a stitch', *Focus*, September 2003, p. 57.

Killer Aspartame and Diet Drinks

Aspartame, the common sweetener in low-calorie diet drinks, has not had an easy run since it was approved by the American Food and Drug Administration in 1981. Today, over 100 million people consume aspartame daily in over 1500 food products. However, the famous 'Nancy Markle email' blames aspartame for some 92 conditions ranging from headaches, fatigue, multiple sclerosis and systemic lupus erythematosis to dizziness, vertigo, diabetes and coma.

Aspartame is produced by combining two common amino acids — phenylalanine and aspartic acid. These amino acids are, like the other 18-or-so common amino acids, found in the proteins we eat and are part of our regular food intake. In aspartame, the phenylalanine has been modified by the addition of a methyl group chemical. The job of the gut is to prepare food so that it can enter the bloodstream. Because the aspartame molecule is too big to get into the bloodstream, the gut breaks it down into three smaller chemicals — phenylalanine, aspartic acid and methanol.

Like all good myths, this one has a germ of truth to it. Under certain circumstances, two of these chemicals (phenylalanine and methanol) can be poisonous.

The first chemical is the natural amino acid, phenylalanine. It is claimed that the phenylalanine is poisonous on the grounds that cans of diet drinks have a health warning: 'Phenylketonurics: Contains Phenylalanine'.

Phenylalanine is in fact toxic to people with phenylketonuria, a very rare disease which affects one in 15 000 people. These people are usually diagnosed soon after birth with the Guthrie 'heel prick' test. In these people, phenylalanine is not broken down and can rise to toxic levels, causing brain damage. Phenylketonurics, placed on a special restricted diet to minimise their intake of phenylalanine, can live normal lives.

There is more phenylalanine in 'regular' foods than in diet drinks. For example, a can of diet drink has 100 mg of phenylalanine, an egg 300 mg, a glass of milk 500 mg and a large hamburger 900 mg. These are foods that phenylketonurics are taught to avoid. However, the other 14 999 people out of every 15 000 don't have to worry about the toxic effects of phenylalanine.

Make mine a . . .

Aspartame, the common sweetener in low-calorie diet drinks, has not had an easy run since the FDA approved it in 1981.

Aspartame is produced by combining two amino acids. These amino acids are, like the other 18-or-so common amino acids, found in the proteins we eat as part of our regular food intake.

Some studies show that diet drinks with artificial sweeteners can stimulate the appetite – thus defeating the whole purpose of the diet drink.

Refreshing, and drain clearing

Killer Diet Drinks

It is claimed that certain 'studies' show that diet drinks containing aspartame can lead to reactions ranging from headaches, fatigue, multiple sclerosis, vertigo and diabetes to coma.

The second chemical is the alcohol called methanol. (There are many chemicals in the alcohol family. Ethanol is the one that is good for us in small quantities — which is amazing for a chemical that can strip stains off a floor, and pickle and perfectly preserve small animals.) It is true that methanol in large doses is toxic. However, a can of diet drink will yield 20 mg of methanol, a very small dose, and easily handled by the body. Like phenylalanine, this chemical is found in our regular diet. A glass of fruit juice will give you 40 mg of methanol, and an alcoholic drink 60–100 mg.

There's one final argument against the toxicity of diet drinks. None of the peer-reviewed medical literature shows a relationship between the consumption of diet drinks, and any of the 92 diseases that aspartame supposedly causes.

And what of Nancy Markle? She has never been found.

Aspartame

Aspartame is a low-calorie sweetener, which was invented in 1965. Weight for weight, it is about 200 times sweeter than sugar. It goes under many names — 'Equal', 'NutraSweet', 'Spoonful', 'E951' etc.

According to the US Food and Drug Administration, the ADI (Acceptable Daily Intake) is about 50 mg/kg of body weight per day. For a 75 kg person, this is the equivalent of 20 cans of diet drink per day.

It has remarkably few adverse reactions. Repeated studies have shown that it does not cause allergic reactions, headaches, cancer, epilepsy, multiple sclerosis, Parkinson's disease, or Alzheimer's disease. It does not affect vision, or cause changes in mood, behaviour or thought processes. It does not increase haemorrhagic risk, and it has no bad effects on dental health. On the other hand, some studies show that diet drinks with artificial sweeteners stimulate the appetite, which can lead to eating when you're not hungry — which defeats the whole purpose of diet drinks.

Even the concern about the effect of heat on aspartame — e.g., when diet drinks are left in direct sunlight — seems ungrounded. No new toxic chemicals are created by the heat. The aspartame simply breaks down, and the taste of the drink becomes less sweet than before.

References

'Kiss My Aspartame', www.snopes.com/toxins/aspartame.asp: Urban legends reference pages.

Zehetner, Anthony & McLean, Mark, 'Aspartame and the Internet', *The Lancet*, vol. 354, 3 July 1999, p. 78.

The Black Box

The Black Box was a long time coming. In the early 1900s, the Wright Brothers invented a primitive device to record the revolutions of the propeller. By the late 1950s, this had evolved into the first flight data recorder, commonly called the Black Box. Whenever there is a plane crash, or a near crash, the highest priority of the accident investigators is to find the Black Box. Which isn't black at all.

Since the 1960s, some 800 aircraft have been destroyed in crashes — and the Black Box has always survived. It came into existence because of the Comet, the first jet airliner. In 1953, Comet jets began to fall out of the sky — and nobody knew why. It took a lot of expensive testing to work out what had happened.

In 1957, an effort to make it easier to find the cause of a crash, the US Civil Aeronautics Authority proposed that all aircraft heavier than 20 000 lbs (about 9 tonnes) should carry a data-recording device. It would capture a few fundamental flight conditions of the plane, such as its direction, speed, altitude, vertical acceleration and time. Since then the technology has evolved from recording information on metal foil and steel wire to recording on magnetic tape. The latest generation has no moving parts, and records directly onto solid state memory.

The single Black Box has evolved into two Black Boxes. One of them is the CVR (Cockpit Voice Recorder) that can record up to two hours of the conversations and sounds in the cockpit. The latest FDR (Flight Data Recorder) can now record 24 hours of information about some 700 different aspects of the plane — such as oil pressure and rotation speed of all the moving parts in each of the engines, the angles of the flaps and the temperatures in the cargo hold. To give the Black Boxes maximum protection, they are located in the part of the plane that is usually last to hit the ground in a crash — the tail. The whole front of the plane is a crumple zone for the Black Boxes.

The latest generation of Black Box is tougher than the wrestlers of the World Wrestling Federation. The solid state memory is surrounded by aluminium, which is enclosed by super-efficient heat insulation material, which is then wrapped in a thick layer of stainless steel. The Black Box has to survive temperatures of 1100°C for one hour, followed by 260°C for 10 hours. In the crash impact test, it has to survive 3400 G of acceleration. Human beings will become unconscious if they experience 5 G for five seconds. The test is usually done by firing

The Black Box (is actually orange)

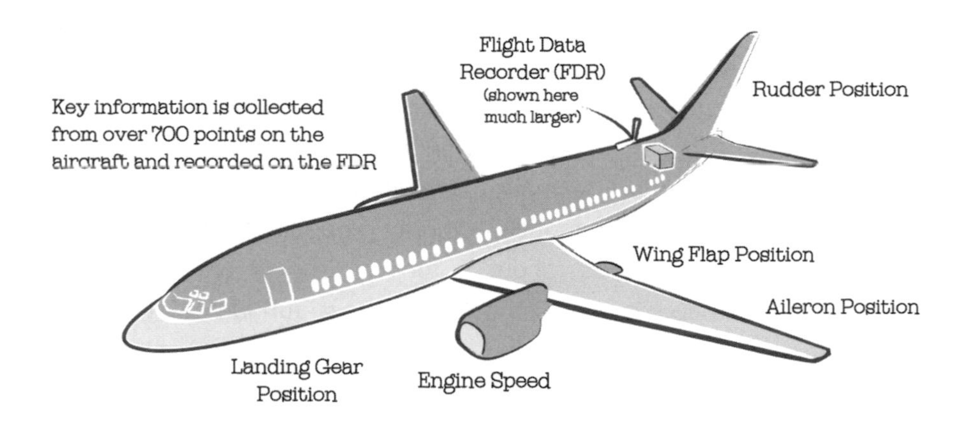

Black Box

One popular theory is that after a decent fire, the 'orange' black boxes are indeed black, from the soot.

the Black Box out of a cannon. In the Pierce test, a 227-kg weight with a hardened steel spike, 6.5 millimetres in diameter, is dropped three metres onto the Black Box. A Black Box can survive the pressure at the bottom of the ocean, and being submerged in salt water for a month. All this high-end engineering means that a Black Box costs about $20 000–30 000.

Of course, the Black Box is not black. In 1965, it was changed to orange — a colour that can be easily spotted. We don't really know why it is called a Black Box. A popular theory is that after a decent fire, the *orange* Black Boxes are indeed black, from the soot.

Why aren't aeroplanes made to withstand the same forces as Black Boxes? First, in a crash, the crew and passengers would still not survive the G-forces. And second, the plane would be too heavy to fly.

Aussie Invention

In 1934, the father of 10-year-old David Warren was killed in one of Australia's first air disasters. His last present to young David was a crystal radio set. David began building radios as a hobby, and so began a life-long interest in electronics.

In 1953, the first Comet crash happened. David was now a principal research scientist at the Aeronautical Research Laboratory (ARL) in Melbourne. He immediately realised that a recording of pilots' voices and various instrument readings would provide invaluable clues as to the cause of the crash. He built the first Black Box, but could not engender any Australian interest in it.

In 1958, he got a lucky break. When the secretary of the UK Air Registration Board, Sir Robert Hardingham, visited ARL, David showed him his unofficial project. Sir Robert recognised its potential immediately and soon David Warren and his first Black Box were on their way to London. Various companies also saw its uses and built their own versions.

Black boxes became compulsory in Australia only after a Fokker Friendship crashed at Mackay in Queensland in 1960. In 1963, Australia became the first country to have a voice recorder in the cockpit.

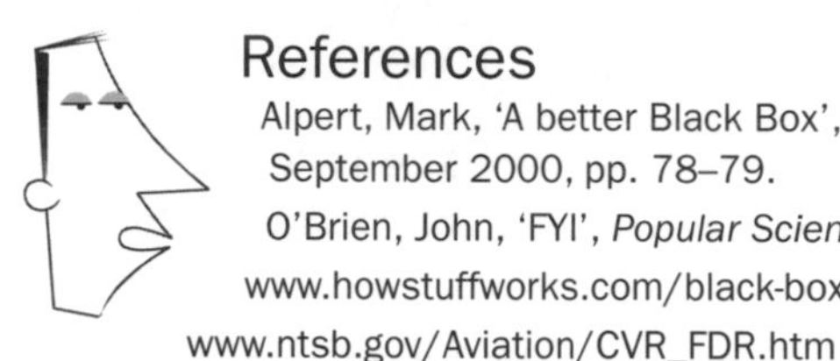

References

Alpert, Mark, 'A better Black Box', *Scientific American*, September 2000, pp. 78–79.

O'Brien, John, 'FYI', *Popular Science*, March 2002, p. 79.

www.howstuffworks.com/black-box.htm

www.ntsb.gov/Aviation/CVR_FDR.htm

Duck Quacks Don't Echo

In 2003, a popular myth — that a duck's quack doesn't echo — was scientifically debunked at Salford University in northwest England, at the annual meeting of the British Association for the Advancement of Science.

I was asked a question about this myth a few years ago on my science talk-back show on Triple J radio. I wasn't able to answer it because I hadn't read any research on the topic. However, on the face of it, it sounded like a ridiculous claim for three reasons.

First, each of the many species of duck has its own different quack. In fact, to make it more complicated, there are gender differences in quacks. For example, the female Mallard duck has a loud honking quack, while the male Mallard duck has a softer, rasping quack.

Second, most species of duck spend much of their time on the water — usually out in the open. Because there are not many hard reflective surfaces near most bodies of water, you don't get many echoes, anyway.

Third, why should the quack of a duck (alone, of all sounds ever made on our planet) have some magical property that makes it echo-free?

Then a Triple J listener rang in with what the Intelligence Community calls 'ground truth'. His family owned a duck farm, and he assured me that the quack of their ducks most certainly did echo off the walls of the sheds.

And this is where my understanding of ducks' quacks rested, until Professor Trevor Cox from the Acoustics Research Centre at the University of Salford reported his research. The focus of his study was Daisy the Duck. There was nothing special about Daisy — Professor Cox just rang local duck farms until he found one (Stockley Farm) that would lend him a willing duck.

Professor Cox has worked with problems in sound for many years. For example, you've probably heard the sound from the loud speakers at a railway station echoing away until it's almost unrecognisable. Professor Cox is the guy who can create a virtual prototype of a railway station inside a computer, and then adjust the design until you can hear the sound clearly. He can do the same for concert halls and even restaurants — in fact his work could save you from having to shout across the restaurant table just to be heard. And he's also worked on using trees to absorb the sound of traffic and aeroplanes.

Being an expert in sound, he knew what to do with Daisy and her quack.

Donald knew something was not quite right at Echo Cliff ... but he just couldn't put his wing on it

No Echo

Professor Trevor Cox from the Acoustics Research Centre at the University of Salford focused his studies on why duck quacks don't echo. After much testing he concluded that the humble quack does in fact echo.

First, he got her to quack in an anechoic chamber — a specially designed room that deadens echoes. Her quack sounded like a regular quack, but a little softer than he expected.

Second, he got Daisy to repeat her quack in a reverberation chamber, which artificially enhances the echoes. The quack did indeed echo — in fact, it sounded rather sinister.

Third, he used a computer to simulate her quack in a concert hall — and sure enough, there was a small echo. And the same happened when he simulated Daisy's quack as she flew in front of a virtual cliff inside his computer.

But he noticed two odd things.

First, Daisy's quack didn't finish sharply (like a hand clap), but tailed off softly. It faded away gently and gradually, making it hard to tell the difference between the original quack and the echo. Second, because her quack was actually quite soft, the echo was even softer. The combination of these two factors means that even if you do get an echo, it is lost in the tail end of the original quack.

Professor Cox is hoping to use this knowledge to improve echo-ridden environments, such as railway stations and restaurants. And I guess that the original myth about the sound of the duck's quack being echo-free is just quackers ...

Ducks Have Accents

According to Dr Victoria de Rijke from Middlesex University ducks have different 'accents', depending on what kind of noise they are subjected to in their local area.

Dr de Rijke lectures in English, with a special interest in phonetics (how language sounds to the ear). She is also the leader of the Quack Project. Her team records the sounds of children with different native languages (e.g. English, Vietnamese, Arabic and Tamil) copying the sounds of common animals, such as ducks.

In an interview with *The Guardian* newspaper, she said, 'London ducks have the stress of city life and a lot of noise to compete with, like sirens, horns, planes and trains. The cockney (London) quack is like a shout and a laugh.' Therefore cockney ducks place their quack energy into the part of the sound spectrum where the background noise is quiet, so that other city ducks can hear them more easily. Dr de Rijke continued, 'On the other hand, the Cornish ducks have a big field to roam in and their quiet surroundings make all of a difference. The Cornish ducks made longer and more relaxed sounds, much more chilled out. They sound like they are giggling. So it is like humans: cockneys have short and open vowels, whereas the Cornish have longer vowels and speak fairly slowly.'

She now plans to study ducks' quacks in Newcastle, Liverpool and Ireland.

References

'I'm a duck, me old cock sparrer', *Sydney Morning Herald*, 5–6 May 2004, p. 19.

Radford, Tim, 'Scientist proves echo claim is just plain quackers', *Sydney Morning Herald*, 18 September 2003.

Switch on Light Bulb

An enduring childhood memory is being told by your parents not to keep turning the electric light switch on and off. Supposedly 'each time you switch the light on, you burn up enough power to run the light bulb for 30 minutes'. And in retaliation, many cheeky kids would deliberately wait until their parents left the room and then flick the switch hundreds of times to send their parents broke.

However, once you start doing the calculations, you realise that this is a ridiculous claim.

Enclosed inside the average light bulb is a thin wire made of silver-coloured metal called tungsten. When the tungsten is cold, it has a very low resistance to electricity. As soon as you flick on the light switch, a huge number of electrons flow through this thin wire which heats up very quickly, and starts to glow — first yellow, and then red and finally white. As it heats up, the resistance increases, reducing the numbers of electrons flowing in the wire. (The number of electrons tells you the current — more electrons means more current.) Within one-fifth of a second of start-up, the light has come up to full brightness, and the electrical current stabilises at a steady value.

When your parents told you to stop flicking the switch on and off, they were probably thinking about this initial surge in electricity. But how big is this surge? Is the number of electrons that flow in this first fifth of a second equal to the number of electrons that flow in 30 minutes of normal running?

Suppose that your parents were indeed correct about the surge of current in the first fifth of a second being equal to the current used in 30 minutes. There are 1800 seconds in 30 minutes. During the fifth-of-a-second start-up, your light bulb would have burnt 1800 x 5 = 9000 times the average power consumption of your light bulb. If you have a 100 watt light bulb, then each time you switch on the light, you burn 900 000 watts, or nearly a megawatt of power. Mind you, that would be for only one-fifth of a second. This is roughly the power output of a very small power station. A sudden drain of this much power would vaporise the wires in your house, and make the lights dim in your suburb or town.

In fact, this initial surge of current would probably burn up one-tenth of a second's worth of regular electric light burning — or perhaps a maximum of one second. On the other hand, your parents were right about so many other things — such as the need to eat lots of vegies, and to have a good night's sleep.

Stop it ... you're wasting power!

Light Switch

It was thought that turning the light on and off used up more electricity than if you just left it on. WRONG. If you think about it, each time it is off, it's using less electricity than if it was on.

History of Light Bulbs

The light bulb could have been invented in 1666 — if somebody had seen the light. This was the year of a giant storm in London. One night, lightning hit St Paul's Cathedral so frequently, that the thick copper straps carrying the electricity down to the ground from the lightning rods on the roof glowed a dull red colour — just like an electric light bulb starting up.

In 1801, Sir Humphry Davy ran electricity through strips of platinum, and they glowed — but not for long. In 1841, Frederick de Moleyns (also in England) generated light by running electricity through powdered charcoal, and obtained the first patent for an incandescent bulb. Again, his bulb didn't last long.

The secret was keeping oxygen (in the air) away from the hot glowing material — and this needed the invention of good vacuum pumps. It was a close race between Sir Joseph Wilson Swan (England, 1878) and Thomas Alva Edison (United States, 1879), who each came up with carbon filaments. But Edison got all the credit, perhaps because he had developed the other technology (e.g. power plant and transmission lines) needed to have a practical lighting system. Today, the bulbs contain an inert gas and the inside of the bulb is not a vacuum.

Faster Light Bulbs

Light bulbs with a faster reaction time are making it safer for car drivers who are braking behind you.

The normal light bulb in the brake-light circuit of your car is very similar to the incandescent light bulb of 1911, when tungsten filaments were introduced. You hit the brake pedal with your foot, making electricity flow into an incandescent light bulb at the rear of your car. After a short delay to heat the tungsten wire to white hot, the bulb reaches full brightness. Only then does the driver behind you realise that you have hit the brakes — and only then do they start the process of hitting their brakes.

A relatively new innovation is the use of the red LED (Light Emitting Diode) in the brake-light circuit. The red LED emits light within a millionth of a second after the electricity hits it. Compared to an incandescent bulb, it has no significant time delay. This means that the person following can see the brake light come on sooner, which at 60 kph, means that they get a braking distance of a few extra metres.

References

Mills, Evan, 'Eleven energy myths: from efficient halogen lights to cleaning refrigerator coils', *Science Beat*, Berkeley Lab, 24 April 2001.

Sydney Morning Herald, Good Weekend/Spectrum, 18 August 2003, p. 22.

www.lbl.gov/Science-Articles/Archive/energy-myths3.html

Cat Years

Many of us own a cat. In fact, about one in every three Australian households owns one. A cat's age is often measured in human terms, that is, one cat year is equal to seven human years. This standard myth might make the maths nice and simple, but the reality is much more complex.

The average life expectancy for the domestic cat is about 14 years, but for the feral cat it's closer to two years (due to road accidents, infectious diseases, poisoning, etc.). The maximum life expectancy of a cat is around 20–22 years, although one domestic cat was reported to have lived for 34 years.

Three main factors affect the maximum life span of an animal. The first factor is intelligence — the smarter the animal, the better it can adapt to its environment, and survive any hazards. The second factor is the environment and its associated hazards. In the wild most animals are killed by accidents or natural predators, well before their agility decreases as a result of ageing. The third factor is nutrition — too much or too little food will shorten the animal's life span.

The Feline Health Center at Cornell University in the United States advise that cats reach the young adult stage at 18–24 months (whereas human beings take about 22 years to reach this stage). Therefore a cat that is one calendar year of age is about 16 cat years of age. Curiously, one more calendar year adds another six cat years — roughly equivalent to 22 human years. And after that, you simply add four cat years for each calendar year. So a four-year-old cat is the equivalent age of a 30-year-old person, a 10-year-old cat the equivalent age of a 54-year-old person, and a 20-year-old cat would be 94 cat years of age.

The scale is roughly the same for dogs, but with one major difference. In general, the larger the dog, the younger it is when it dies. For example, a Great Dane would be old at nine calendar years of age, while a Chihuahua would be considered 'old' at 15 calendar years of age.

However, for both cats and dogs, the simple calculation — one calendar year equals seven pet years — is definitely a mythconception. Think about it. There are many 12-year-old cats that can jump tall fences and chase other cats — but you won't see similar behaviour in most 84-year-old human beings.

Are you a pussy with time?

The domestic cat

A human time piece

Cat Years

A cat's age is often measured in human terms, that is, one cat year is equal to seven human years. The truth is, we equate cat years to human years to make it easier for us when they pass away. We can say they led a good and full life.

Egyptian Cats

In ancient Egypt the cat had great religious significance. The ancient Egyptians mummified dead cats and placed them in tombs.

In the 19th century, many amateur archaeologists were running wild across Egypt. They found so many dead cats in the tombs, that they just threw them away. The mummified cats were used for fertiliser and as ballast on ships.

Recent x-rays of these cats found that most of them did not die of old age. In fact, these mummified cats were only a few months of age, and were all in good health until they were strangled. The theory is that the cats were bred by priests in the temples, and then killed and mummified, to be sold to temple visitors or tourists as votive offerings.

Of course there were different grades of mummies — a well-wrapped ones fetched a higher price.

References

Brace, James J., 'Theories of ageing: an overview', *Veterinary Clinics of North America: Small Animal Practice*, November 1981, pp. 811–814.

Thrusfield, M.V., 'Demographic characteristics of the canine and feline populations of the UK in 1986', *Journal of Small Animal Practice*, 1989, pp. 76–80.

No Lead in the Pencil

We have been using lead, a grey or silver-white soft metal, for thousands of years. Although lead has many useful properties, it is, unfortunately, also quite toxic to human beings. As a result, lead has been replaced, in many applications, by less toxic metals. Even so, many people still believe the mythconception that lead pencils contain lead.

For tens of thousands of years, our ancestors drew on cave walls with pieces of charcoal or sticks. About 3500 years ago, during the 18th Dynasty in Egypt, the technology had advanced from burnt sticks to a thin paint brush around 15–20 cm long. The brush made a fine, wet dark line. About 1500 years later, the Greeks and the Romans realised that a sharpened lump of lead would mark papyrus with a dry, light line.

Another 1500 years later, during the Middle Ages, European merchants commonly used a metal stylus (called a 'metalpoint') which could make faint marks on paper. The merchants made these faint marks more visible by first coating the surface with a material like chalk. If the metal were lead — it usually was — it would mark the fingers, so it was often wrapped in paper, string or wood.

The beauty of the modern pencil is that it combines the best qualities of the paint brush and the lump of metal in one product. The modern pencil makes a line that is very useful because it is both dry (so that it doesn't run) and dark (so that it's easy to see).

...not enough lead in your pencil, eh?

The 'Lead' Pencil

The modern pencil makes a line that is very useful because it is both dry (so that it doesn't run) and dark (so that it's easy to see).

Recipe for success

It wasn't until 1795 that Nicolas-Jacques Conté worked out a method of converting low-quality graphite into fine writing material. He ground low-quality graphite very finely, mixed it with clay, fired the mixture at high temperature and added wax.

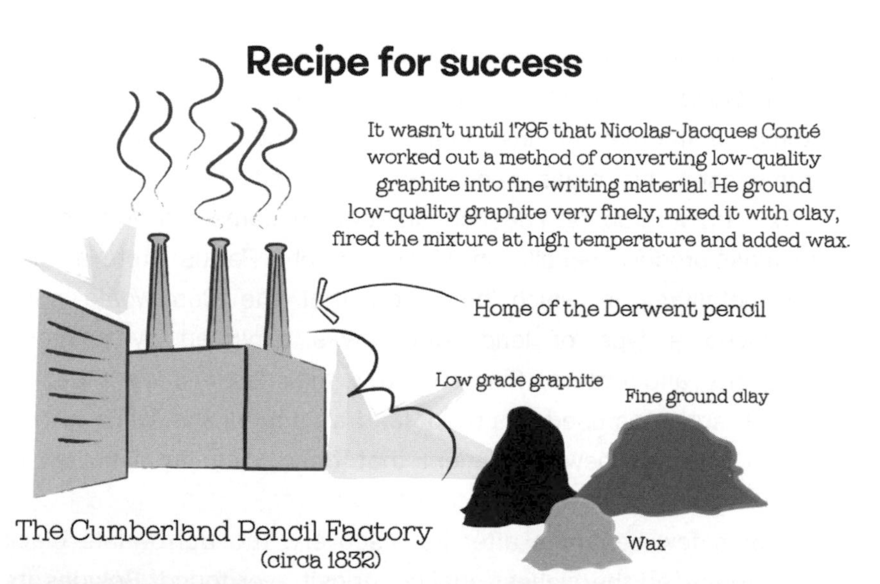

The Cumberland Pencil Factory (circa 1832)

The Lead Pencil

The modern 'lead' pencil is very good technology. It is entirely self-contained, uses no messy liquids such as ink, can produce a continuous line for some 35 kms, makes well-defined marks that are relatively smudge-proof and is easy to erase.

The modern lead-free lead pencil first appeared in the early 1500s, in Borrowdale, in the Cumberland Lakes District of England. Legend has it that when a large tree blew over, the local shepherds noticed a black material clinging to the roots. They tried to burn it, thinking that it was coal — but it would not burn. However, they quickly found a use for it — marking their sheep.

The shepherds had discovered graphite, which is actually a variety of carbon. But at the time they thought it was just a variety of lead, so they called it 'black lead'.

We know that black lead was not commonly used in pencils up to 1540, because in that year, the Italian writing master, Giovanbattista Palatino, wrote a book describing what he thought would be 'all the tools that a good scribe must have'. It did not include anything that looked like a pencil, or contained graphite. But 25 years later, in 1565, Konrad Gesner, a Swiss naturalist and physician, wrote a book on fossils. In his book, he describes and makes a drawing of a new writing instrument that seems to be the first primitive black-lead pencil. Lead pencils were now becoming common.

In 1609, a character in Ben Jonson's play *Epicoene* describes some mathematical instruments including 'his square, his compasses, his brass pens, and black-lead to draw maps'. In 1622, in Nuremberg, Friedrich Staedtler became the first person to mass-produce pencils. In 1683, Sir John Pettus wrote a book on metallurgy in which he noted that the Borrowdale mine produced a type of lead, which was exploited by painters, surgeons and writers. Painters drew their preliminary sketches with it, surgeons used this black lead as a medicine, while writers rejoiced in this new instrument that freed them from having to carry a bottle of ink.

For a few centuries after its discovery, the Borrowdale black lead remained the highest quality deposit ever found. Besides its medical, painting and writing applications, graphite had very important strategic military functions in casting cannon balls and other metal objects. Therefore, on 26 March 1752, the House of Commons passed a bill entitled, 'An Act for the More Effectual

Securing Mines of Black Lead from Theft and Robbery'. This Act made it a felony, punishable by hard labour and/or transportation to the colonies, to steal this high quality graphite.

For many years the English would not allow their enemies to use the pure Borrowdale graphite. It was not until 1795, through the urging of Napoleon, that Nicolas-Jacques Conté finally worked out a method of converting low quality graphite into a fine writing material. He ground low quality graphite very finely, mixed it with finely ground clay, fired the mixture at high temperatures, and finished by adding wax before inserting it into slim wooden cases. In 1832, a pencil factory started operations near the Borrowdale graphite supply. In 1916, it became the Cumberland Pencil Factory, which produced the Derwent pencils still loved by school children.

However, although writing pencils made of graphite were first used around 1565, writing pencils that used lead were still very commonly used in the 18th century. Why? Because they were cheaper, even if they were toxic. You certainly wouldn't want to suck on a 'lead' pencil if it really had lead in it. In fact, lead pencils became extinct only in the early 20th century.

The modern lead pencil is very good technology. It is entirely self-contained, uses no messy liquids such as ink, can write a continuous line for some 35 km, makes a well-defined mark that is relatively smudge-proof and is easy to erase.

Today we have glasses made of plastic, tins made of aluminium and golfing irons made of titanium. So it really shouldn't bother us that lead pencils use graphite.

Forms of Carbon

It was only in 1779, that the Swedish chemist K.W. Scheele proved that the Borrowdale black lead was not lead, but in fact a form of carbon. It was given a new name, 'graphite', which comes from the Greek verb *graphein*, 'to write'.

Graphite is a variety of carbon which is the sixth lightest element, fitting between boron and nitrogen. It is not very common in the Earth's crust (making up 0.025% by weight), but it makes more chemical compounds than any other element.

There are three forms of pure carbon — when it exists as pure carbon, and is not combined with any other element.

Diamond, which is the hardest element known, is made up of carbon atoms arranged in a series of tiny pyramids.

Another form is called 'buckyballs', where the carbon atoms (typically 60, but there can be more or fewer than 60) are arranged in a hollow ball, like a soccer ball. 'Buckytubes' or 'nanotubes' are very similar — and are carbon atoms arranged in hollow tubes, like a drinking straw. Buckyballs and buckytubes are very strong.

In graphite, the carbon atoms are arranged in circles of six that are joined, side by side, to make thin sheets. These thin sheets are stacked on top of each other. The chemical bonds are very strong within each sheet — but are very weak from one sheet to the next. This weak joining makes the sheets easy to slide over each other, making graphite an excellent lubricant. The weak joining also means that when you wipe graphite over a slightly rough surface like paper, a few of the sheets rub off, leaving a mark on the paper. Graphite is also one of the softest known minerals.

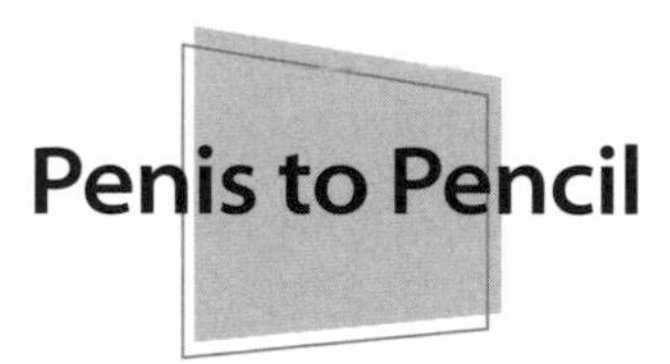

Penis to Pencil

The word 'pencil' comes from the Latin word *penicillum*, which was a collection of fine animal tail hairs that had been shoved into a hollow reed. It got its name from *peniculus*, which was the Latin word for 'brush'. In turn, the 'brush' got its name from the Latin word *penis*, which meant 'tail' — the location on the animal from which you plucked the hairs.

References

'Graphite, Lead, Conrad Gesner,' *Encyclopaedia Britannica*, (DVD), © 2004.

'How the lead gets into a pencil', *How Is It Done?*, Readers Digest Association, London, UK, 1990, p. 13.

Binney, Ruth, *The Origins of Everyday Things*, Readers Digest Association, London, UK, 1998, p. 223.

Petroski, Henry, *The Pencil: A History of Design and Circumstance,* Alfred A. Knopf, New York, 1989.

Milk Makes Mucus

There are many myths about the cow, an important economic animal in our society. One of the more popular myths is that 'milk makes mucus'. This usually means that if you have a cold, and then drink milk (or take other milk products), your nose will start generating huge quantities of lovely green mucus. But there doesn't seem to be any link between milk and mucus — apart from the fact that they both begin with the letter 'm'.

Mucus is a viscous liquid, that is, it is fairly thick and slow-moving — like honey. It usually wets, lubricates and protects many of the 'pipes' in our body that are related to moving food, air, urine and sexual juices around. Mucus consists of water, a special protein called mucin, dead white blood cells, cells that have been shed from the surface of the local 'pipe', various chemicals from the immune system (such as antibodies) and inorganic salts.

There are several types of mucous glands in different locations around the body, and they each make their own specialised type of mucus. In the stomach, a thick layer of mucus stops the acids from digesting the stomach. (After all, if the acids can dissolve meat, what stops them from dissolving the stomach itself? The answer is mucus.) In the mouth, mucus stops the moist, inside surfaces from drying out, and also helps the food slide down. In the nose, mucus is an important part of our air-conditioning system. Not only does the mucus system trap bacteria, small particles and dust, but it also brings the incoming air up to around 100% humidity by the time it hits the back of the throat.

And when you have a cold, the mucous-producing cells in the nose make more mucus.

Dr Carole Pinnock and her colleagues from the Royal Adelaide Hospital in South Australia explored the widely-held belief that 'milk makes mucus'. They started with 60 healthy volunteers aged 18–25 years, whose milk consumption ranged from zero to 11 glasses of milk per day. The investigators squirted Rhinovirus-2 (one of the many viruses that can cause the common cold) up the noses of the volunteers and supplied them with tissues and plastic bags to collect the 'nasal debris'. They found that the weight of mucus dribbled and blown out of the nose varied between zero and 30.4 g, and that the maximum mucous production happened on the third day after infection.

Dr Pinnock also found there to be absolutely no relationship between how much milk the volunteers drank, and how much mucus they produced. Interestingly, the volunteers who believed

The Milky Way

When you have a cold, the mucus-producing cells in the nose make more mucus. Dr Carole Pinnock from the Royal Adelaide Hospital in South Australia ran a study and found that there was absolutely NO relationship between how much milk we drank, and how much mucus we produce.

MILK

SNOTTY NOSE

Milk Makes Mucus

This usually means that if you have a cold, and then drink milk (or consume milk products), your nose will begin generating huge quantities of green mucus. But there doesn't seem to be any link between milk and mucus...apart from both beginning with 'm'.

that 'milk makes mucus' claimed that they coughed more often and were more congested. However they did not produce any more nasal mucus than their more sceptical fellow volunteers.

Some studies have shown that milk products are unlikely to 'have a specific bronchoconstrictor effect in most patients with asthma' — in other words, milk will not close up the airways. And other studies showed that milk does not cause 'acute or delayed asthmatic symptoms or deterioration of pulmonary function'.

Milk has many benefits. In Western societies, milk and other dairy products supply up to 50% of vitamin A needed by young children and adults, over 50% of the calcium, 33% of the riboflavin, and 20% of the protein, vitamin B12 and retinol. So, if you don't have a true 'milk allergy', denying yourself dairy products can compromise your nutritional status.

Why do so many people believe that milk causes mucus? Probably for the simple reason that it looks a little like mucus.

References

Low, P. P., Rutherfurd, K.J., Gill, H.S. & Cross, M., 'Effect of dietary whey protein concentrate on primary and secondary antibody responses in immunized BALB/c mice', *International Immunopharmacol*, March 2003, pp. 393–401.

Pinnock, C.B., Graham, N.M., Mylvaganam, A. & Douglas, R.M., 'Relationship between milk intake and mucus production in adult volunteers challenged with rhinovirus-2', *American Review of Respiratory Diseases*, February 1990, pp. 352–356.

Tampon Tampering

In 1998, I got my first (of many) emails about the dangerous products in tampons. Tampon manufacturers were accused of deliberately including asbestos in tampons to produce more bleeding, thus increasing tampon sales. It's easy to understand why women would be concerned about the quality of their tampons. After all, they use them internally, and most women have heard of Toxic Shock Syndrome (although it is incredibly rare).

The source of this information in the email was said to be 'a woman getting her PhD at the University of Colorado'. This person was never identified.

Although the email has been slightly modified over the years it keeps resurfacing. Nowadays I don't hear the asbestos claim too frequently, but there are plenty of other claims levelled at tampons.

People often ring my Triple J science talk-back show asking if tampons are made from 'dangerous' rayon, or laced with cancer-causing dioxins. One version of the email claims that rayon 'is a highly absorbent substance, and therefore when fibres are left behind in the vagina, it creates a breeding ground for the dioxin'. This is ridiculous. Dioxins are a family of chemicals. They are not living creatures, and cannot breed.

And what about the claim that rayon sheds fibres? When you run cotton or rayon T-shirts through your clothes dryer, you'll find a lot more lint collected after the cotton load than after a rayon load. And anyway, on the spectrum between 'natural fabric' and 'synthetic fabric', rayon is much closer to the 'natural fabric' end. Rayon is made of cellulose fibres which come from wood pulp — so you can't really hit it with the 'synthetic is bad and natural is good' equation.

The dioxin claim has (like all good myths) a tiny element of truth in it. There are some dioxins in tampons. But there are dioxins virtually everywhere in our environment — what matters is the quantity. It turns out that there are higher levels of dioxins in your body, than there are in tampons. In 1995, the United States FDA (Food & Drug Administration) measured dioxin levels in cotton and rayon used to make tampons. The dioxin levels ranged from 'non-detectable to one part in three trillion'. To put it in plain English, one part in three trillion is roughly one teaspoonful in a lake 10 metres deep and one kilometre square. This is less than the exposure to dioxins that you get from the activities of daily living. Today, Johnson & Johnson bleach their tampons, using a process that does not produce dioxins.

Tampon tampering

The source of this information is 'a woman getting her PhD at the University of Colorado'.

In short:

1. Rayon is NOT synthetic, it is derived from tree cellulose fibres.

2. Rayon sheds less fibres than cotton.

3. There are SOME dioxins in tampons. BUT there are dioxins virtually everywhere in our environment – what matters is the quantity.

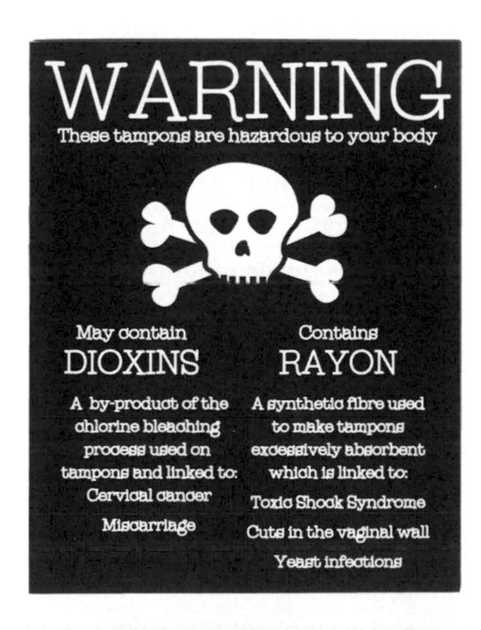

Tampon Tampering

The claim: Tampon manufacturers have been accused of deliberately including asbestos in tampons to produce more bleeding. This was written by 'a woman getting her PhD at the University of Colorado'.

The 'inventor' of the tampon, Dr Earle Cleveland Haas, was selected in 1969 by the London *Sunday Times* as one of the '1000 Makers of the 20th Century'. In 1929, Dr Haas, from Denver Colorado, tried to relieve the discomfort that his female patients suffered while wearing external pads during menstruation. He eventually devised a cotton cylinder that could be inserted into the vagina with a disposable applicator. He sold the patents and trademark to Gertrude Tenderich, who was the first president of the Tampax Sales Corporation, which was established on 2 January 1934.

And Dr Haas? He kept on trying to improve tampons, right up until he died in 1981, at the age of 96.

The Email

Disclaimer from Dr Karl: This is an edited version of the notorious 'tampon' email. As the email travels around the word, people sometimes change it slightly before passing it on. The following version covers the essential parts of the email.

Dear Friends: I am passing this along because this information may improve your life and I care about you.

Have you heard that tampon makers include asbestos in tampons? Why would they do this? Because asbestos makes you bleed more, and if you bleed more, you're going to need to use more. Why isn't this against the law, since asbestos is so dangerous? No wonder so many women in the world suffer from cervical cancer and womb tumours.

A woman getting her PhD at the University of Colorado at Boulder sent the following:

'I am writing this because women are not being informed about the dangers of something most of us use — tampons. I am taking a class this month and I have been learning a lot about biology and women, including much about feminine hygiene. Recently we have learned that tampons are actually dangerous.

HERE IS THE SCOOP: Tampons contain two things that are potentially harmful: rayon (for absorbency) and dioxin (a chemical used in bleaching the products). The tampon industry is convinced that we, as women, need bleached white products in order to view the product as pure and clean. The problem here is that the dioxin produced in this bleaching process can lead to very harmful problems for a woman. Dioxin is potentially carcinogenic (cancer-associated) and is toxic to the immune and reproductive systems. It has also been linked to endometriosis and lower sperm counts for men — for both, it breaks down the immune system.

Rayon contributes to the danger of tampons and dioxin because it is a highly absorbent substance.

Therefore, when fibres from the tampons are left behind in the vagina (as it usually occurs), it creates a breeding ground for the dioxin.'

References

Haas, Dr Earle, US Patent no. 1 926 900, 12 September 1933.

Mikkelson, Barbara, 'Asbestos in tampons', Urban legends homepage: www.snopes.com/toxins/tampon.htm

Hindenburg and Hydrogen

In the mid-1930s, if you could afford to fly across the Atlantic Ocean, there were two choices — noisy, small and cramped aeroplanes, or quiet and spacious airships that got their lift from huge bladders filled with hydrogen gas. Back then, it was still an even bet as to which technology would not become obsolete — the faster and noisier aeroplane, or the slower and more relaxed lighter-than-air airship.

Following a disastrous event in 1937, the aeroplane became the favoured technology. The enormous hydrogen-filled Nazi airship, the Hindenburg, was slowly manoeuvring in to dock at a 50-metre high mast at the Lakehurst Air Base, in New Jersey. This was the Hindenburg's twenty-first crossing of the Atlantic Ocean. Suddenly, there was a spark on the Hindenburg, and then flames. Newsreel film crews captured the sudden disaster as the Hindenburg burst into enormous plumes of red-yellow flames, and collapsed to the ground. Over 30 of the 97 people on board died. The disaster was blamed on the extreme flammability of the hydrogen lifting gas that filled most of the airship.

Hydrogen's reputation (as extremely flammable) still troubles car manufacturers today, as they explore the use of hydrogen as a safe, non-polluting alternative to fossil fuels for powering cars. But it turns out that the extreme flammability of hydrogen is a mythconception.

The Hindenburg was the largest aircraft ever to fly — longer than three football fields (about 250 m long). It was powered by

That looks like a large ...

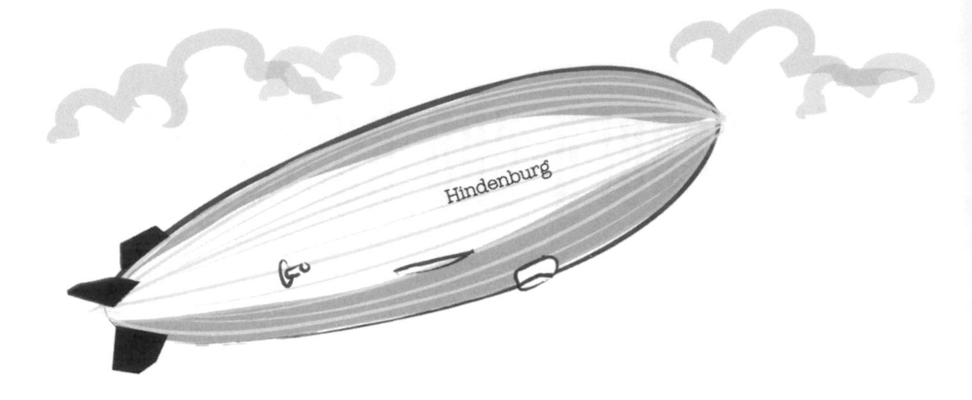

The Hindenburg was the largest aircraft ever to fly . . . longer than three football fields (250 m long).

For once, the gas had nothing to do with it!

The belly of the beast

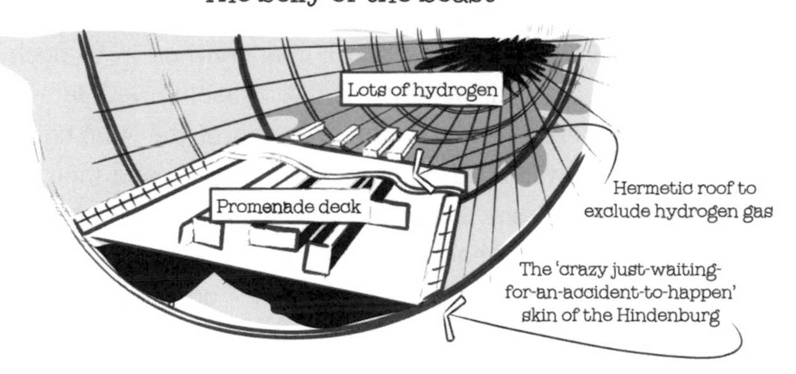

The Hindenburg was covered in cotton fabric that had been waterproofed. To achieve this, the fabric had been swabbed with cellulose acetate (very flammable) and then covered with aluminium powder (which is now used as rocket fuel).

Hindenburg Myth

The Hindenburg burned with a red flame. But hydrogen burns with an almost invisible bluish flame. The hydrogen was totally innocent and the disaster was not caused by a hydrogen explosion.

four enormous 1200-HP V-16 Mercedes-Benz diesel engines that spun six-metre wooden propellers. It cruised at 125 kph (faster than ocean liners and trains), and when fully loaded with fuel, had a range of some 16 000 km. It was opulently furnished and almost decadently luxurious — each of the 50 cabins had both a shower and a bath, as well as electric lights and a telephone. The clubroom had an aluminium piano. The public rooms were large and decorated in the style of a luxury ship — and the windows could be opened. It was a little slower than the aeroplanes of the day — but it was a lot more comfortable.

The Hindenburg was painted with silvery powdered aluminium, to better show off the giant Nazi swastikas on the tail section. When it flew over cities, the on-board loudspeakers broadcast Nazi propaganda announcements, and the crew dropped thousands of small Nazi flags for the school children below. This is not surprising, because the Nazi Minister of Propaganda funded the Hindenburg.

At that time, the US government controlled the only significant supplies of helium (a nonflammable lifting gas), and refused to supply it to the Nazi government. So the Hindenburg had to use flammable hydrogen.

As the Hindenburg came in to land at Lakehurst on 6 May 1937, there was a storm brewing, and the enormous amount of static electricity in the air charged up the aircraft. When the crew dropped the mooring ropes to the ground, the static electricity was earthed, which set off sparks on the Hindenburg.

The Hindenburg was covered with a cotton fabric, that had been waterproofed. To achieve this, the fabric had been swabbed with cellulose acetate (which happened to be very flammable) and then covered with aluminium powder (which nowadays is used as rocket fuel to propel space shuttles into orbit). Indeed, the aluminium powder consisted of tiny flakes, which made them very susceptible to sparking. It was inevitable that a charged atmosphere would ignite the flammable skin of the airship.

The Hindenburg burned with a red flame. But hydrogen burns with an almost invisible bluish flame. In the Hindenburg disaster, as soon as the hydrogen bladders were opened by the flames, the hydrogen inside would have escaped up and away from the burning airship — and would not have contributed to the ensuing fire. The hydrogen was totally innocent. In fact, in 1935, a helium-filled airship with an acetate-aluminium skin burned near Point Sur in California with equal ferocity. The Hindenburg disaster was not caused by a hydrogen explosion.

The lesson is obvious — the next time you build an airship, don't paint the inflammable acetate skin with aluminium rocket fuel.

Hydrogen

Hydrogen is the most abundant element in the Universe — about 75% of all the mass in the Universe. But it's only the ninth most abundant element on Earth, and makes up just under 1% of the mass of our planet.

Hydrogen is an odourless, colourless and tasteless gas and is also the simplest and lightest chemical element.

It seems that Paracelsus, the 16th century German-Swiss alchemist and doctor, may have handled hydrogen. He discovered that when he dissolved a metal in acid, it produced a gas that would burn. In 1766, the English chemist Henry Cavendish went one step further with this gas, which was then called 'inflammable air' or 'phlogiston'. He actually measured the amount of gas that he got from a certain amount of acid and metal, and even measured its density. In 1776, J. Waltire noticed that when he burnt hydrogen, he also made some droplets of water. It was the French chemist Antoine-Laurent Lavoisier who came up with the name 'hydrogen', from the Greek, meaning 'water generator'.

Liquid hydrogen is used as a rocket fuel, and when it burns with oxygen it produces temperatures of around 2600ºC. Hydrogen was once used to fill balloons, but is now mostly used to make ammonia and methanol, to remove sulphur from petrol, and to make food products such as margarine.

References

'Hindenburg burns in Lakehurst crash: 21 known dead, 12 missing; 64 escape', *New York Times*, 6 May 1937, p. 1.

Lemley, Brad, 'Lovin' hydrogen', *Discover*, November 2001, pp. 53–58.

Antiperspirant and Cancer

In Australia, breast cancer will affect more than one in every 12 women. Over the past two decades, thanks to improved diagnostic methods, the reported incidence of breast cancer has been increasing. A popular rumour claims that the increase in breast cancer is caused by the increasing use of antiperspirants — and this old mythconception received a new burst of life with the advent of the Internet and email.

Each armpit has about 25 000 sweat glands, which can produce about 1.5 ml of sweat every 10 minutes. Sweat is mostly salt water, with microscopic amounts of various chemicals. Each square centimetre of your armpit has about one million bacteria. They use the sweat to reproduce themselves and release waste products — some of which are the dreaded smells of body odour. Most antiperspirants use some kind of aluminium chemical, which seems to work by turning into an insoluble aluminium hydroxide gel inside the sweat glands. This gel physically blocks the sweat from getting out, and therefore stops the production of bacteria and body odour.

Part of the 'science' behind this antiperspirant/cancer myth is the claim made in a circulating email that 'the human body has a few areas that it uses to purge toxins: behind the knees, behind the ears, groin area and armpits. The toxins are purged in the form of perspiration.' This is totally incorrect. Actually, it is the kidneys that get rid of unwanted metabolic by-products (the toxins). In fact, the main purpose of perspiration is to act as a cooling mechanism.

The email goes on to say that because the toxins can't escape via the armpits, they lodge in the lymph nodes between the armpits and the upper outer quadrant of the female breast — which is the part of the breast closest to the lymph nodes. '... this causes a high concentration of toxins and leads to cell mutations: aka *cancer*.' Pathologists around the world have examined millions of lymph nodes from breast cancer patients — and while they have seen cancer cells that have travelled from the breast to the lymph node, they have not found high concentrations of these toxins.

The email also claims that the cancer then spreads from the lymph nodes to the breast and that 'nearly all breast cancer tumours occur in the upper outside quadrant of the breast area'. Wrong again. First, lymph node fluid flows in the opposite direction — from the breast to the lymph nodes. Second, the real figures are that about 60% of breast cancers happen in the upper outer

Antiperspirant and Cancer

It is claimed that the use of antiperspirants can lead to cases of breast cancer.
Part of the claim is that the toxins being purged from the lymph nodes can't escape via the antiperspirant-filled-armpit. This supposedly causes high concentrations of toxins that lead to cancer.

Actually, the flow of fluids happens in the opposite direction – from the breast to the lymph nodes.

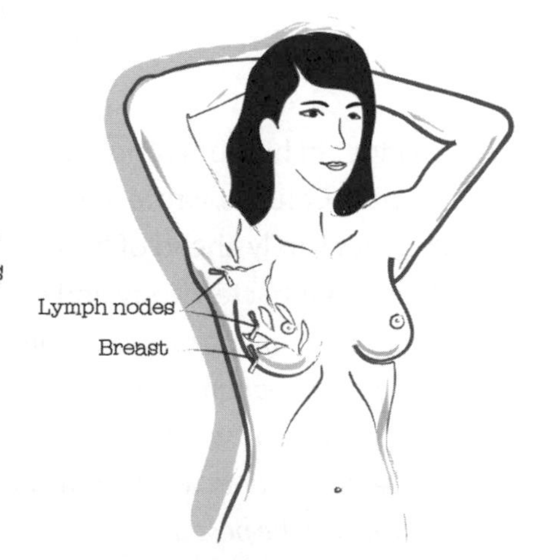

Antiperspirant

A popular rumour claims that the increase in breast cancer is caused by the increasing use of antiperspirants. Of course this old mythconception received a new burst of life with the advent of the internet and email.

quadrant. This is probably because that's where 60% of the total breast tissue is.

Dana K. Mirick and her colleagues addressed this myth in their paper 'Antiperspirant use and the risk of breast cancer', in the *Journal of the National Cancer Institute*, in October 2002. They compared 813 patients — diagnosed with breast cancer between November 1992 and March 1995 — with 793 healthy women (who were matched with the patients by five-year age groups). They found that the risk for breast cancer was not increased by any combination of using an antiperspirant, a deodorant or shaving.

The curious thing about the antiperspirant/cancer (and other 'causes' of cancer) myths is that they focus on factors that play little or no role in cancer. At the same time, they ignore known risky activities such as alcohol (linked to breast cancer) and smoking (linked to lung cancer, bladder cancer, kidney cancer and other cancers of the head and neck).

Armpit Sweat

Ben Selinger, in his book, *Chemistry in the Marketplace*, relates that the Roman poet Catullus, wrote a poem about armpit smell in 50 BC. Catullus was surely ahead of his time, because he seems to know an awful lot about bacteriology and molecular biology. The poem runs:

An ugly rumour harms your reputation,

Underneath your arms they say you keep a fierce goat which alarms all comers — and no wonder,

For the least Beauty would never bed with rank beast,

So either kill the pest that makes the stink,

Or else stop wondering why the women shrink.

The bacteria that live in your armpits produce chemicals. One of them, 4-ethyloctanoic acid, does have a goat-like smell. But this chemical is especially attractive to female goats on heat. So if your armpits generate this smell, be selective when you lift your arms, or you might end up with a lot of lust-crazed female goats chasing you.

References

'Cancer myth dispelled', *New York Post*, 16 October 2002, p. 9.

'Dangerous personal care products?' *Choice*, March 2004, pp. 24–28.

'Study: Deodorants don't cause cancer', *USA Today*, 16 October 2002, p. 5.

Mirick, Dana K., Davis, Scott & Thomas, David B., 'Antiperspirant use and the risk of breast cancer', *Journal of The National Cancer Institute*, vol. 94, no. 20, pp. 1578–1580.

Selinger, Ben, *Chemistry in the Marketplace*, Harcourt Brace & Company, Australia, 1998, pp. 61, 119–121, 272.

Dinosaurs and Cave People

Most people have a sneaking fascination for dinosaurs, those enormous lumbering beasts from days gone by. There are many dinosaur myths, but the most popular of the top ten dinosaur myths (according to *New Scientist* magazine) is that human beings lived at the same time as the dinosaurs.

Dinosaurs first evolved around 228 million years ago, and survived for a very long time, to around 65 million years ago. (As a comparison, human beings have been around for only three million years.) Dinosaurs were different from modern reptiles in that they had five vertebrae in their hips rather than two. Their limbs were also directly underneath them, rather than sprawled out to the sides, allowing them to run faster. The first known fossil discoveries were made in Si Chuan in China around 300 BC, when Chang Qu wrote about 'dragon bones' — which were actually dinosaur bones.

The modern era of dinosaur fascination began in the 1820s when the clergyman, William Buckland, and the physician, Gideon Mantell, separately discovered some strange, enormous bones in quarries in southern England. In 1842, the English anatomist Richard Owen proposed that these giant extinct animals should be called 'dinosaurs', from the Greek words meaning 'fearful lizard'. The flood gates of dinosaur fascination were well and truly opened.

The first verified movie dinosaur had its guest appearance in D.W. Griffith's 1912 film, *Man's Genesis*. But the first dinosaur movie to catch the imagination was *Gertie the Dinosaur*, an animated film of 1914. Since then, hundreds of movies featuring dinosaurs have been made, including the wonderful 1966 offering of *One Million Years B.C.*, in which Raquel Welch — complete with a fetching fur bikini, waxed armpits, and shining and manageable hair — coexists with dinosaurs. Dinosaurs and human beings also coexisted in the cartoon series *The Flintstones*, and in the *Dinotopia* series of books by James Gurney. In the 1920s, the *Pellucidar* series of books by Edgar Rice Burroughs fed the myth by describing underground caverns on our planet where dinosaurs and cave people coexisted. You can also find human beings coexisting with the dinosaurs in books of knitting patterns, and in the teachings of many different Christian fundamentalist religious groups who believe that the world is about 6000 years old.

This is nothing but fantasy. Geology has many separate, accurate and reliable dating techniques. They all confirm that dinosaurs had pretty well died out 65 million years ago. Human

Cave Folk versus Dinosaurs

Time Difference

Dinosaurs had pretty well died out 65 million years ago. Human beings didn't really begin to evolve into their current form until a few million years ago.

beings on the other hand didn't really begin to evolve into their current form until a few million years ago.

But in this case, science gives us a romantic way out to believe in dinosaur/human coexistence. Birds, which evolved about 150 million years ago, can be counted as dinosaurs. So if we didn't live with dinosaurs, at least we lived with their immediate descendants, who now have feathers.

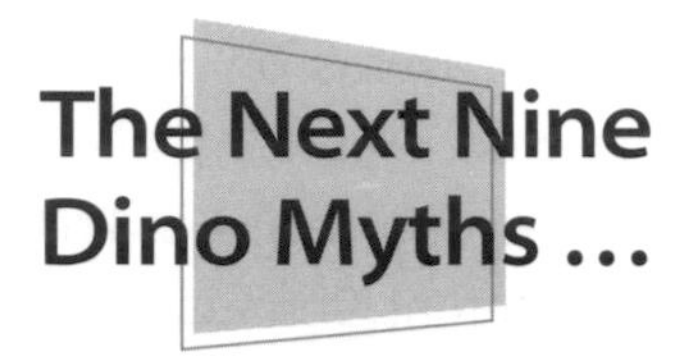

The Next Nine Dino Myths ...

And here are the next nine most popular myths according to *New Scientist* magazine:

- Mammals appeared on the planet only after the dinosaurs died out.
- Dinosaurs died out because mammals ate their eggs.
- Dinosaurs were wiped out by a single event — a big asteroid which hit the Earth 65 million years ago, near what is today called the Gulf of Mexico.
- Dinosaurs died out because they were unsuccessful in evolutionary terms.
- All the dinosaurs died out 65 million years ago.
- All the large reptiles from prehistoric times were dinosaurs.
- Marine reptiles such as the plesiosaur and the ichthyosaur were a variety of dinosaur.
- Flying reptiles were dinosaurs.
- Dinosaurs were slow and sluggish animals.

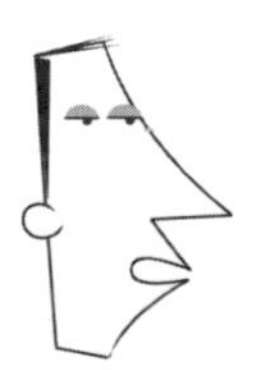

References

Torok, Simon, *Wow! Amazing Science Facts and Trivia*, ABC Books, Sydney, Australia, 1999, p. 78.

Encyclopaedia Britannica, (DVD), © 2004.

Growth Spurts

It is generally believed that children grow bigger gradually, until their growth rate tapers off to zero. Medical books show the growth curves of babies as smooth curves, without any jumps. However, scientists who actually measured the heights of babies showed that babies spend most of the time not growing at all, and then have an overnight growth spurt!

For some time now we have known that newborn babies lose weight immediately after birth, and then regain it all within about 10 days. At birth, the average baby is about 50 cm long, and will grow another 25–30 cm during the first year of life. Generally, babies double their birth weight by five months of age, and triple it by the time they reach one year of age.

Then, their rate of growth slows down, and during the second year of life, they gain another 2.5 kg, and grow about 12 cm taller. During the third, fourth and fifth years, their growth is relatively constant — they gain 2.5 kg and grow about 7 cm taller each year.

However, not everyone believed that this growth happened evenly throughout the whole year. So a team consisting of Dr Lampi from the University of Pennsylvania, and Drs Veldhuis and Johnson from the University of Virginia Health Sciences Center, set out to investigate.

They tracked 31 healthy Caucasian infants (19 girls and 12 boys), between the ages of three days and 21 months, measuring the height of these infants at varying time intervals.

For the first 21 months of life, 10 of these infants were measured once a week, 18 were measured twice a week, while three were measured every single day. ('Three' isn't a very big sample size, but it's a start.)

The team went to a lot of trouble to make sure that their measurements were as accurate as possible. They used a special measuring device with a fixed headboard, and a moveable footboard. One of the specially-trained observers would hold the infant's head, while the second observer would apply gentle but firm pressure to the infant's body to make sure that the legs were straight, and that the feet were at right angles to the legs. The footboard was then brought into contact with the infant's feet, and a final check was made on the proper alignment of the infant's head and body. If everything was correct, the height was then read off the scale on the device, to the nearest half millimetre.

The team also made 90% of all measurements within three hours of the same time on each visit. This was to account for the

From size 000 to size 00, overnight

The 'test' baby before bed

The 'test' baby next day
Note increased size due to growth spurt.

Baby Growth

It was generally thought that baby growth happened gradually. However, scientists who actually measured the lengths of babies found that babies spend most of the time not growing at all, and then suddenly have an overnight growth spurt.

fact that you are tallest first thing in the morning, and then gradually shrink during the day. To be extra sure, the team would then make a duplicate measurement within one hour — and the same people would always do the measurements.

The results were astonishing. When they measured the infants weekly, they found that some of them did not grow at all for up to 63 days (that's over two months), after which the infant would have a sudden growth spurt of up to 25 mm. And in the infants measured daily, they found periods of non-growth varying between 2 and 28 days, with growth spurts of between 8 and 16 mm. The top end of the range — 16 mm — is a huge amount to grow overnight. (These individual jumps in height were between 2.5 and 10 times greater than any error in the measurement of height.)

The team are not sure just why this pattern of long periods of non-growth, followed by a sudden large overnight growth spurt, should happen. But it does reassure parents, who swear that their children have grown out of their clothes in just a few days. And it could help you answer the old question of why your children get sudden intense unexplained pains in various parts of their bodies. If you put it down to 'growing pains', you will probably be right.

TV Stunts Growth?

It seems, from one study at least, that too much television can stop teenage boys from growing.

The Menzies Research Institute in Hobart looked at 130 boys, aged 16 and 17 years, for a period of six weeks during winter. If the boys watched TV for less than one hour a day, they grew 7.5 mm. If they watched TV for two to three hours each day, they grew a lesser amount — only 2.5 mm. And if they watched more than four hours of TV, they stopped growing entirely during the six-week window.

What's going on?

Professor Graeme Jones thinks that the key is vitamin D, which is necessary for bones to grow. Children need about eight hours of sunlight each week on the hands and feet to supply their vitamin D requirements. If they watch lots of TV, they don't get the sunlight exposure. They could get their vitamin D by eating deep-sea fish, such as salmon — but unfortunately, most Australian teenage boys don't eat lots of salmon.

Will the teenagers ever recover the lost growth? At this stage of research, nobody knows.

Professor Jones limits his own children to just one hour of TV each day and encourages them to spend time outdoors in the winter time.

References

'Growth and development', *Nelson Textbook Of Pediatrics*, USA, 1987, pp. 6–25.

'US study reveals surprise baby growth', *Sun-Herald*, Sydney, 1 November 1992, p. 9.

Elliott, Dr Elizabeth, 'Out-of-date birth charts revamped', *Medical Observer,* 21 February 1997, p. 22.

Heinrichs, Claudine, Munson, Pete J. & Counts, Debra R., 'Patterns of human growth', *Science*, vol. 268, 21 April 1995, pp. 442–446.

Lampi, M., Veldhuis, J.D. & Johnson, M.L., 'Saltation and stasis: a model of human growth', *Science*, vol. 258, 30 October 1992, pp. 801–803.

Truth Serum

Since the terrorist attacks on the twin towers of the World Trade Center on 11 September 2001, there have been calls for shortcuts to extract information from 'uncooperative sources'. Sir James Stephens wrote in 1883 that '... it is far pleasanter to sit comfortably in the shade rubbing red pepper into some poor devil's eyes than to go about in the sun hunting up evidence.' Most of us would consider this an act of torture, and be appalled by it. Instead, we might prefer to use the painless and well-known 'truth serum' to get the information we want.

But there's a catch. There is no drug known that will relax a person's defences so that they will happily and reliably reveal the truth, the whole truth, and nothing but the truth.

The small core of truth in this myth is that some drugs will loosen one's tongue. In 77 AD, Pliny the Elder wrote, '*In vino, veritas*', or 'In wine, there is truth'. In the early 1900s obstetricians were using scopolamine — a plant extract — to produce a state of relaxation and 'twilight sleep' during childbirth. (It has also been used in small quantities to relieve motion sickness.) The doctors soon noticed that the women would sometimes speak extremely candidly under the influence of the scopolamine.

In 1922, Robert House, an obstetrician from Dallas, Texas, used scopolamine on supposedly guilty prisoners. He was convinced that the then-current police administration was corrupt and he wanted to free the innocent. Under the influence of this drug, the prisoners claimed to be innocent and the subsequent trials agreed. Dr House enthusiastically claimed that scopolamine created a state where a person 'cannot create a lie ... and there is no power to think or reason'. This was very exciting stuff and knowledge of his work spread rapidly. The *Los Angeles Record* was probably the first newspaper to use the catchy phrase 'truth serum'.

Soon psychiatrists were using scopolamine and barbiturates (such as sodium pentothal) in an attempt to shine light on the hidden recesses of the mind. Both the barbiturates and the scopolamine seemed to work best when the subject was in the sedative stage of anaesthesia where one's regular thought patterns were disrupted.

It took a few decades to realise that the trouble with all the 'truth serums' was that a subject's delusions, fears and fantasies could not be distinguished from reality. People would even confess

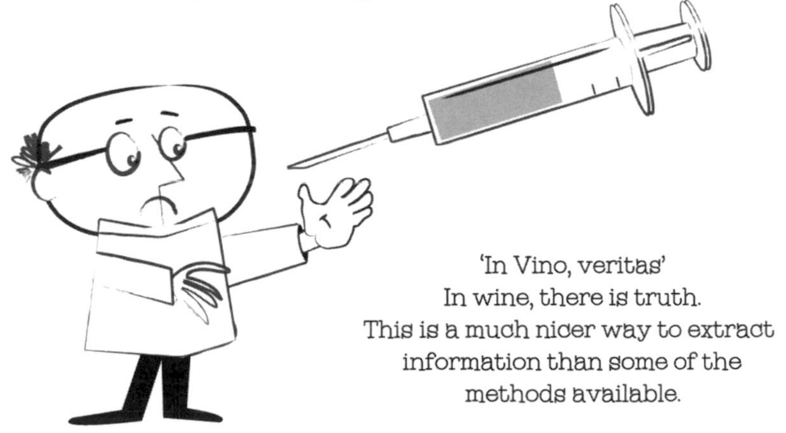

Truth Serum

There is no known drug that will relax a person's defences so that they will happily and reliably reveal the truth, the whole truth, and nothing but the truth.

to crimes that they clearly did not commit, perhaps even the murder of a non-existent stepmother. They would be more likely to confess to something if they had a strong desire to be punished.

Various 'secret government' organisations have long been interested in a 'truth serum'. So the CIA (the United States Central Intelligence Agency) ran the MKULTRA Project — a 25-year program of research into behavioural modification. They found that probably the most effective aspect of the so-called truth serums, is that they falsely led the subject to believe that they had revealed more than they actually did. This often tricked the suspect into later revealing the truth.

All that a truth serum probably does is loosen a person's tongue — but there's no guarantee of truth, only a distorted peek into somebody's subconscious ...

How to Give 'Truth Serum'

Over the years various drugs have been tried as 'truth serums'. They include scopolamine, the barbiturates (sodium pentothal) and sodium amytal, and the hallucinogenics (LSD and psilocybin). The administering of the drug was critical.

For example, four distinct stages of the descent into unconsciousness can be described:

Stage 1: Sedative, relaxed stage.
Stage 2: Unconscious, with exaggerated reflexes.
Stage 3: Unconscious, with no reflexes.
Stage 4: Death (oops, gone too far!).

Stage 1 can be further broken down into three planes:

Plane 1: Zero or small sedative effect.
Plane 2: Cloudy, fuzzy, calm, no memory of this plane upon waking.
Plane 3: Poor coordination, slurred speech, suggestible.

Truth serums are most effective at loosening tongues during Plane 3 of Stage 1. This can involve many injections over a period of 2–10 hours.

So even the single shot of truth serum, shown in the movies, is inaccurate!

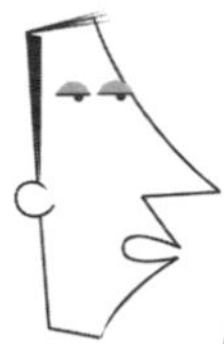

References

'Use drug on al-Qaeda prisoners: ex-CIA chief', *Sydney Morning Herald*, 27–28 April 2002, p. 13.

'Project MKULTRA, the CIA's Program Of Research In Behavioral Modification', 1977 *Senate Hearing on MKULTRA: 'Truth' Drugs in Interrogation*.

Nails and Hair Grow After Death

Thomas Decker, the 17th-century English dramatist believed that people's hair continued to grow for a while after they died. He wrote eloquently: 'Hair is the robe which curious nature weaves to hang upon the head, and to adorn our bodies. When we were born, God doth bestow that garment. When we die, then like a soft and silken canopy it still is over us. In spite of death, our hair grows in the grave, and that alone looks fresh, when all our other beauty is gone.'

In the powerful novel, *All Quiet on the Western Front*, the author Erich Maria Remarque imagines that the nails of a buried dead friend continue to grow in a spiral. And even today, in our supposed Days of Enlightenment, many of us still believe that both our hair and our nails continue to grow after we die. (Perhaps it's the morbid fascination that we have for moody and melancholic facts ...)

Hair can have many functions — many animals use it as insulation against the cold, while other animals use it as camouflage, allowing them to fade into the background. Of course, hair is important in sexual attraction and recognition, and even as a sensory organ for nocturnal animals. As far as hair is concerned, human beings are probably the most hairless of mammals.

Hair grows at about 13 millimetres per month. Fingernails grow at about a quarter the speed of hair — roughly one millimetre per week in young people, and one millimetre per 10 days in older people. Toenails grow more slowly again — at about half the speed of fingernails.

Hair and fingernails do not continue to grow after death.

It's an optical illusion — like when you sit in a stationary train, and the train on the neighbouring track moves forward. For a brief instant, you have the impression that your train is going backwards. (It was Albert Einstein who supposedly said, 'When does Clapham Junction arrive at this train?') In the same way, hair and nails do not advance after death, but instead, the skin retreats.

After we die, we lose water and dehydrate. To compensate for this, funeral homes will try to stop the exposed skin from

I could just die for a trim right now

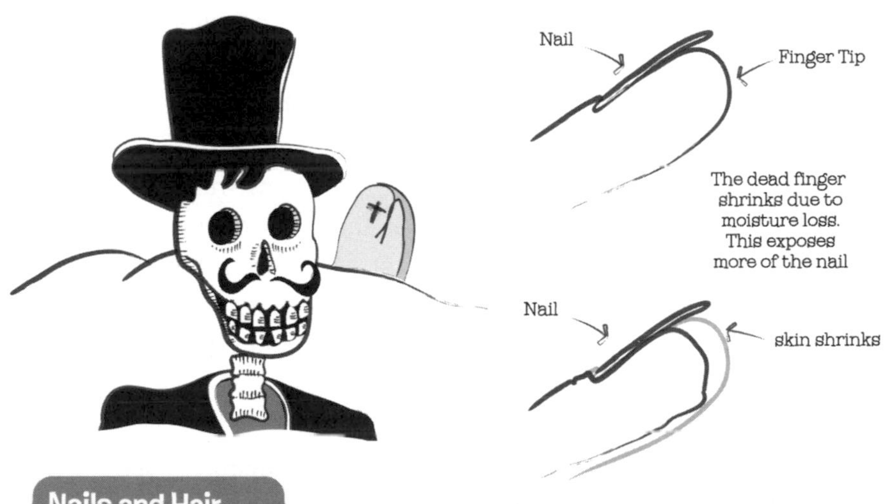

Nails and Hair

Hair and nails don't keep growing after death. The body, which is largely made up of water, begins to dry out. As your skin shrinks around your skull and skeleton, more of the fingernails and hair are exposed giving the illusion that they are growing.

dehydrating by applying various moisturising creams — especially on men who have thick beards.

CSI Hair and Nails

Whenever an attacker struggles with a victim, stuff gets swapped between them. Fragments of skin, strands of hair and clothing fibres can be caught under their nails. Crime scene investigators have long been interested in these materials.

One of the very first papers on this topic was published in France in 1857. Within half a century, it was common to microscopically examine any hair found at a crime scene. In 1931, Professor John Glaister published the book, *Hairs of Mammalia from the Medico-Legal Aspect*.

Because hair is made of proteins that are very resistant to decay, they can survive for a long time after a person's death. During life, hair is very 'absorbent', and will, for example, pick up arsenic in a case of arsenic poisoning.

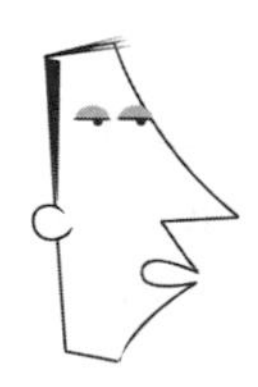

References

'Coffin Nails', www.snopes.com/science/nailgrow.htm: Urban legends reference pages.

Encyclopaedia Britannica, (DVD), © 2004.

Man on Moon Conspiracy

On 15 February 2001, the American Fox TV Network broadcast a program called *Conspiracy Theory: Did We Land on the Moon?* Mitch Pileggi, an *X-Files* actor, hosted this hour-long show, which claimed that NASA had faked the entire Apollo moon project by filming it in a movie studio. This myth has a small following — according to both a 1995 Time Poll and a 1999 Gallup Poll, about 6% of Americans 'doubt' that 12 astronauts walked on the Moon.

The conspiracy theorists cite all kinds of evidence.

For example, they point out that in all the photographs that supposedly show the astronauts on the airless surface of the Moon, you cannot see the stars in the black sky. The explanation is simple. Even today's best quality film cannot simultaneously show both a very bright object (white spacesuit in sunlight) and a very faint object (star). Story Mugrave, an astronaut who has flown in a space shuttle six times, said that whenever he was outside the shuttle in the bright sunlight, he couldn't see the stars either. But when the shuttle was in the shadow of the Earth and his eyes had time to adapt to the darker environment, he could then see the stars. (By the way, all the Moon missions happened during the Moon's day — which lasts about 14 Earth days — so that the astronauts could see what they were doing.) And anyhow, when did you last see stars in the sky in daytime?

The hoax believers also point out that in the photos, the shadows of the astronauts and the various pieces of scientific apparatus on the Moon's surface are not quite parallel. They should be parallel, these doubters claim, if lit by only a single, distant light source such as the Sun. This is true — but only if you are working with both a level surface and a three-dimensional field. When you try to show the three-dimensional reality of a bumpy surface in a flat two-dimensional photograph, the shadows fall in slightly different directions.

The conspiracy theorists also claim that the ripple in the American flag, as seen in the still photos, is proof that the landing was faked in a movie studio, because only moving air can make a flag ripple. This is nonsense — for a few reasons. First, there is no wind in a movie studio — unless the wind machine is switched on. Second, if there were enough wind in a movie studio to ripple the flag, it would have also moved the dust at their feet. But third, and most importantly, the ripple was a well-documented accident. The workshops at the Manned Spacecraft Center in Houston, Texas, attached the nylon American flag to vertical and horizontal bars.

Houston, we have a (lighting) problem

Moon Conspiracy

According to both a 1995 Time Poll and a 1999 Gallup Poll, approximately 6% of Americans 'doubt' that 12 astronauts walked on the Moon.

These bars were telescopic, to save space before they were used. Neil Armstrong and Buzz Aldrin had trouble with the horizontal telescopic rod, and were unable to pull it all the way out. This gave the flag a ripple. Because the flag 'looked' realistic, later Apollo crews intentionally left the horizontal rod partially retracted.

In fact, the wobbling flag helps prove that the astronauts were on the Moon. The flag is wobbling because it has just been set up. And it continues to wobble for a short while in a very unusual fashion. This is because the gravity on the Moon is one-sixth the gravity on Earth, and because there is no air on the Moon to quickly stifle the movement of the flag.

But the incontrovertible proof that human beings did go to the Moon is the existence of a total of 382 kg of Moon rocks, which have been examined by thousands of independent geologists around the world. These rocks have been compared to a few dozen Moon rocks that landed in Antarctica, after being blasted off the Moon by meteor impacts, and to some Moon rocks recovered by unmanned Russian spacecraft. All of these Moon rocks share the same characteristics.

Moon rocks are very odd. First, they have a very low water content. Second, they are riddled with strange little holes, because they have been hit by cosmic rays on the airless surface of the Moon for millions of years. The Moon rocks are very different from Earth rocks, and could not be faked by any current technology. To manufacture fake Moon rocks, you would have to squash them using about 1000 atmospheres of pressure, while keeping them at about 1100°C for a few years. Then, while keeping them under pressure, you would have to cool them slowly for a few more years.

There is another proof. Since 1969, new geological dating methods have been invented, and applied to the Moon rocks — and all the dating methods give the same dates for the Moon rocks. If there was a conspiracy, NASA scientists in 1969 would have to have worked out what new dating methods would be invented over the next 30 years, and fake their rocks accordingly.

After looking at all the evidence, I prefer to follow the words of the 1937 Nobel Prize winner in Medicine, Albert Szent-Gyorgyi: 'The

Apollo flights demand that the word "impossible" be struck from the scientific dictionary. They are the greatest encouragement for the human spirit.'

More Objections

There are dozens of problems with this 'faked moon landing' conspiracy theory, but I will deal with just a few. (If you want to read more, check out Phil Plait's 'Bad Astronomy' home page at www.badastronomy.com).

One problem — how do you fool the entire worldwide network of 400 000 scientists, engineers, clerks, lawyers, accountants, technicians and librarians, who helped to make this monumental project happen?

Another problem — the pictures. NASA broadcast the lunar landings live, and made them available to the TV networks of the world at no charge. Most of these pictures are fairly fuzzy, because the technology wasn't very good in those days. But they also released 1359 ultra-high-quality 70 mm film frames, 17 very high-quality pairs of 35 mm lunar surface stereoscopic photographs, and 58 134 high-quality16 mm film frames. Is this the act of an organisation trying to cover up a big conspiracy?

Real Conspiracy Theory

Why are there no photographs of Neil Armstrong walking on the Moon? On the first Moon mission, Michael Collins stayed in orbit around the Moon, while Neil Armstrong and Buzz Aldrin explored the surface of the Moon.

Here is a neat, and totally unprovable, conspiracy (which I heard from a physicist, who knew another physicist, who had met Wernher von Braun, the famous rocket scientist — so it must be true!).

Apparently, Buzz Aldrin was supposed to be the first man to walk on the Moon. But at the last minute, Neil Armstrong pulled rank — he was the commander of Apollo 11, after all — and decided that he would be the first person to walk on the Moon.

So (according to this conspiracy theory) Buzz Aldrin got his revenge by refusing to take any photos of Neil Armstrong. The only photos of Neil Armstrong on the Moon are tiny reflections of him (taken by himself) in the golden faceplate of Buzz Aldrin's spacesuit helmet.

References

'Apollo Moon Landing — A resource for understanding the hoax claims: did man really walk on the moon?', National Space Centre, UK: www.spacecentre.co.uk.

Matthews, Robert & Allen, Marcus, 'Hot debate: did America go to the moon?', *Focus*, February 2003, pp. 73–76.

Camel Hump

There are two myths about the camel — that it is a horse designed by a committee, and that it can store water in its hump.

Camels originated on the North American continent about 40 million years ago, but have since died out there. About one million years ago, camels reached South America and Asia. During this time, they evolved very nicely to survive in the harshest environments. Their soft feet spread widely to walk easily on snow or sand. They have horny pads on the chest and knees to carry their weight when they kneel down. They are superbly adapted to deal with dust storms — they have hair around their ear holes, two rows of eyelashes and can close their nostrils.

Two species of camel still survive today.

The Arabian (Dromedary) Camel stands about 2 m high at the shoulder, has one hump, and can lope along at speeds of up to 16 kph for 18 hours.

The Bactrian Camel which has two humps has long been a 'beast of burden' in the highlands of central Asia. It has shorter legs than the Arabian Camel, but has a heavier build. It is much slower too (at 3–5 kph), but can cover 50 km a day with a very heavy load.

There are about 17 million Arabian Camels in Africa and the Middle East, and another two million in Pakistan and India. There are only two million Bactrian Camels, mostly in the highlands of central Asia.

What's inside the hump?

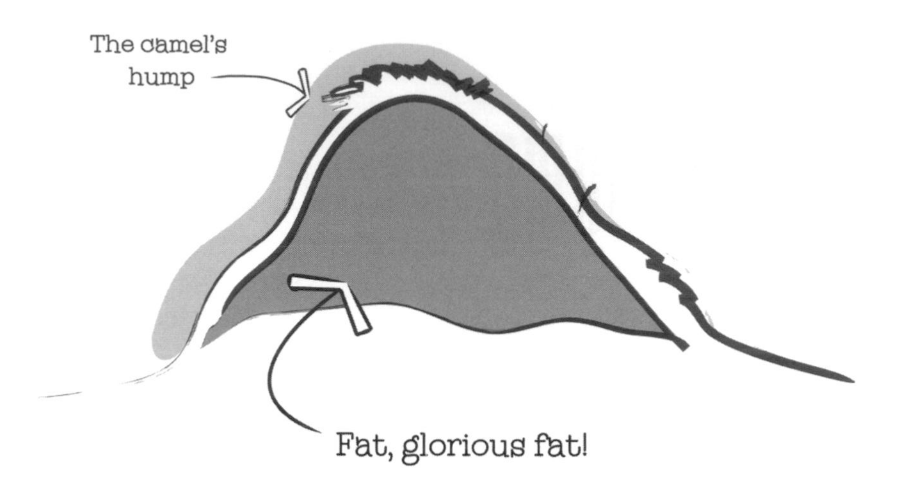

One hump or two ?

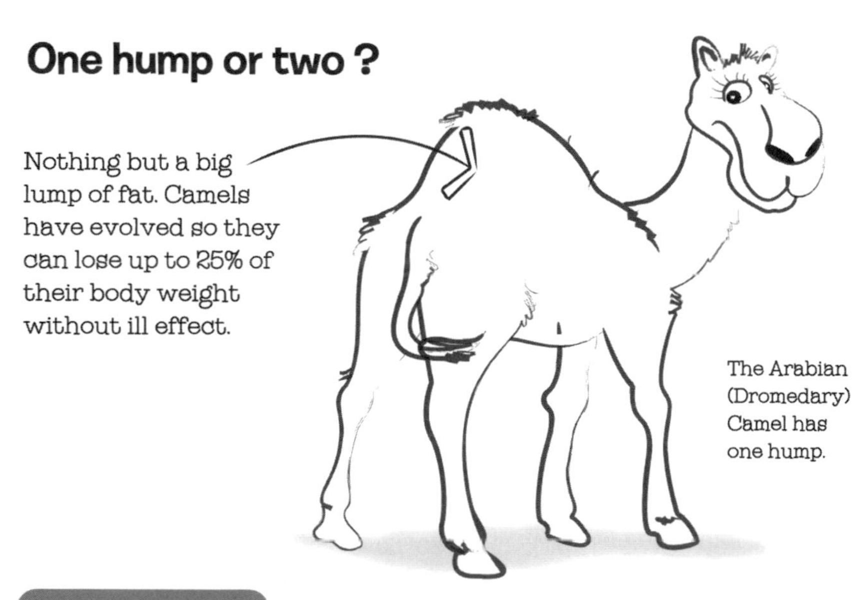

Camel Hump

A camel doesn't store water in its hump. However, it can store up to 35 kg of pure unadulterated fat.

Camels do not store water in their humps. Instead, the hump is used to store most of the camel's fat — up to 35 kg. (This explains why the meat from the rest of the camel is quite low in fat.) One advantage of having so much fat stored on the camel's back is that it insulates the camel from the heat of the desert sun. The fat also acts as an insulator in another way. Fat usually has a very poor blood supply. So even if the fat does get warm, it can transfer only a very small amount of heat to the total blood circulation of the body.

The camel's hump is predominantly a massive food reserve, not a water store. The price it has to pay to access this potential food, is that it loses lots of water vapour from its lungs. When a camel burns a gram of fat for energy, it produces more than a gram of water. But to do this, the camel burns up a lot of oxygen which requires lots of extra breathing — and as it breathes out, it loses much of its water.

How then can a camel go without water for 17 days and survive?

First, camels have evolved so that they can lose up to 25% of their body weight without ill effect. Human beings can't do this easily.

Second, they have evolved in ways to conserve water. They lose very little water in their urine, because their kidneys have a massive concentrating ability, so their urine is like syrup. Their faeces contain so little water that when the round dry pellets pop out, they can be burned as fuel. Camels adjust their body temperature to suit the environment (ranging from 34–41.7°C), and rarely have to waste water by sweating. And when they need more water, they remove it from everywhere in their body except the bloodstream. Their body dehydrates, but the blood circulates normally and does not 'clog up'. No other mammal can do this.

Finally, when they do get near water, there's no horsing around. They can drink as much as a hundred litres in 10 minutes. Their gut can release this very large volume of water very slowly, so that their internal metabolism is not suddenly overloaded.

Camels do not carry water in their hump — and if they were designed by a committee, the members did a very good job.

Camels in Australia

The Ghan is a train that runs between Darwin and Adelaide. The 2979-km line, completed in 2004, was the biggest infrastructure project in Australia in the past half-century. The train is about half a kilometre long, and covers the distance in about two days (allowing for stops and scheduled tourist breaks). It takes about 2 km to bring such a huge train to a complete halt from its top speed of around 100 kph. It also takes about $700 worth of diesel fuel just to get the engine started and stabilised, and fully integrated (electricity, vacuum brakes, etc.) with all the carriages.

The Ghan gets its name from the Afghani camel-handlers who were brought to Australia (along with their camels) 150 years ago. Camels were the ideal way to travel across the deserts of the outback, and the Afghani people had been handling camels for thousands of years. The early explorers, and then the workers who laid the various telegraph lines, loved these sturdy animals that coped easily with the very difficult conditions. Today about 25 000 camels run wild in central Australia.

References

Torok, Simon, *Wow! Amazing Science Facts and Trivia*, ABC Books, Sydney, Australia, 1999, p. 79.

Encyclopaedia Britannica, (DVD), © 2004.

Gun Silencer

An essential part of any thriller movie is the villain screwing a small cigar-shaped silencer onto the end of the barrel of an enormous handgun, and then shooting a few innocent Good Guys in cold blood. The silencer softens the enormously loud report of a firearm to the gentle 'plop' of a stone falling into water. Artistic licence wins again.

First, a little chemistry and physics. A firearm works by setting off a fast-burning chemical, such as gunpowder. The small volume of gunpowder gets converted into a huge volume of gas, which tries to escape. The gas expands, pushing the bullet ahead of it along the barrel of the gun. As soon as the bullet leaves the barrel, a sudden high-pressure wave of gas hits the atmosphere, and slows down. The pressure is quite high — up around 300 atmospheres (about 3000 tonnes per square metre). For comparison, the pressure in a car tyre is about 2–3 atmospheres.

All firearms make at least one sound — when the gunpowder explodes.

The first sound is that of the expanding high-pressure gas. You hear a much more civilised and quieter version of this sound whenever you pop a champagne cork. A silencer can quieten down this first sound. (By the way, the proper name for a 'silencer' is a 'sound suppressor' or 'sound moderator'.)

When the bullet travels faster than the speed of sound there is

Keep it down, I'm trying to kill here ...

Your basic weapon of destruction with silencer attached

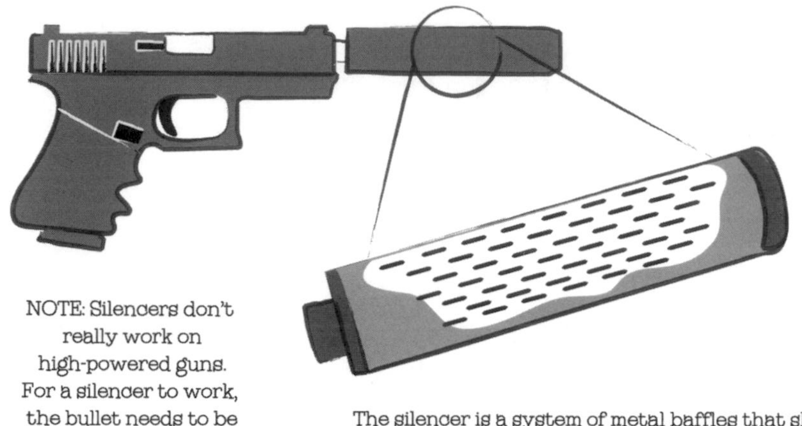

NOTE: Silencers don't really work on high-powered guns. For a silencer to work, the bullet needs to be travelling at sub-sonic speeds.

The silencer is a system of metal baffles that slow down the gases from the exiting bullet. This reduces the level of sound.

Gases hit the baffles and weave their way around while escaping. As they weave, they lose speed and heat up, and in turn, they become quieter.

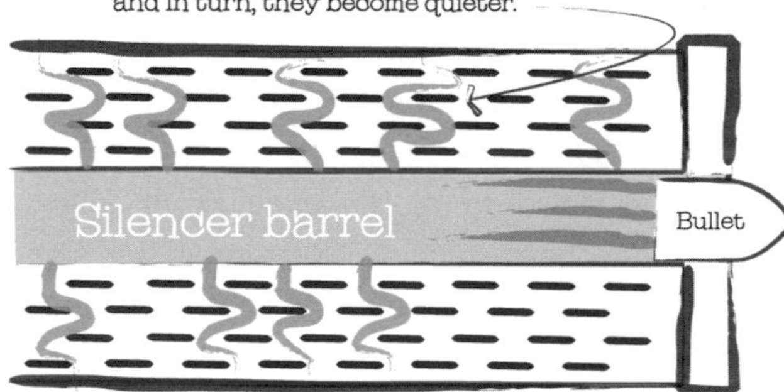

The external surface of the silencer is very hot after use.

Gun Silencers

It has long been believed that gun silencers make a quiet muffled squeal (thanks to movies). A silencer gun will still appear as being quite loud ... something akin to a car door slamming.

a second sound — a genuine mini sonic boom. A silencer can do absolutely nothing to quieten down the sonic boom. The bullet from a high-powered rifle travels faster than sound, giving off a sonic boom. On the other hand, the bullet from a small pistol moves slower than sound, and does not have a sonic boom.

In 1910, Hiram P. Maxim (son of Hiram S. Maxim, who invented the first practical machine gun) patented one of the first usable silencers. Very quickly, people began using silencers for target shooting and crime.

There are a few different designs of silencer, but all of them try to damp down the sudden 'spike' of high-pressure gas from about 300 atmospheres to about 4 atmospheres. If a regular gun shot is like pricking a balloon with a pin and getting a loud sound, a silencer is more like letting the air out slowly through the mouth of the balloon. A typical silencer is like a mini car muffler. The volume inside an effective silencer is about 20 times greater than the volume of the barrel of the gun. Expansion chambers within the silencer allow the gas to expand and slow down, and wire mesh absorbs the heat and interferes with the pressure waves. And of course, the bigger the silencer, the more effective it is in quietening down the bang.

Suppose that the villain has a sensible small gun (which is not as visually exciting as a huge gun), loaded with quieter sub-sonic ammunition — and suppose that the villain uses a realistic large silencer. Even this combination will quieten the 'bang' down to only about 50 dB — which can still be heard about 150 m away in a quiet street! The report doesn't sound like a 'plop' or a 'phut', but more like a car door slamming, or a muffled 'crack' sound.

So the little 'plops' from the silenced guns in the movies probably come from the same sound library that gives us the sound of squealing tyres on bitumen, when the car chase happens on a dirt road.

Who Uses Them?

In real life, gun silencers are used to cull animals in parks, airports, reservoirs, managed forests, etc., mainly as a public relations exercise, so that the general public won't know what is happening.

Tactical response groups who have to enter a building against armed resistance will often use silencers. The sound from firearms inside a building can cause temporary or permanent deafness. Earmuffs will dampen the sound of the firearms — but unfortunately, will also hide any sounds made by the Bad Guys. Therefore police officers will often prefer to use silencers. Silencers have the public relations advantage again here.

Finally, police snipers will sometimes use silencers. To some degree, a silencer will mask the location of the sniper from the target — and again, there's the public relations advantage.

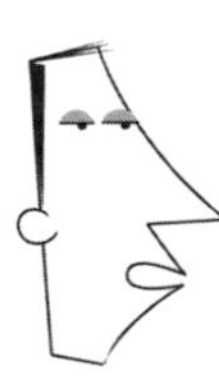

References

Huebner, Siegfried F., *Silencers for Hand Firearms*, Pallandin Press, USA, 1976.

National Rifle Association of America, *NRA Firearms Fact Book*, 3rd edition, USA, 1989.

Knuckle Cracking and Arthritis

There are many old wives' tales including the one that says cracking your knuckles will give you arthritis in your fingers. Because medicine tends to concentrate on the diseases that kill you, not a lot of research has been done on this relatively benign topic. But overall, it seems that cracking your knuckles doesn't make them arthritic.

By the way, 'cracking' a knuckle involves pulling suddenly, and fairly violently, on the end of a finger. If you do it 'correctly', you generate a popping noise. Biomedical scientists have worked out what actually happens when you crack a finger joint by using a sensitive microphone (to listen to, and analyse, the sound) and a strain gauge (to measure the amount of pull on the finger).

The scientists discovered that two separate sounds are actually generated when you crack a knuckle.

The knuckle is the joint that lets your finger bend. In the joint space between the bones is a liquid, and ligaments on each side of this space hold the bone together.

When you pull on your finger (to 'crack' the joint), you make the joint space bigger. As a result, the pressure inside the joint space drops. For a brief instant, the ligaments get sucked inwards. As

Arthritis shmarthritis ... cracking is tough!

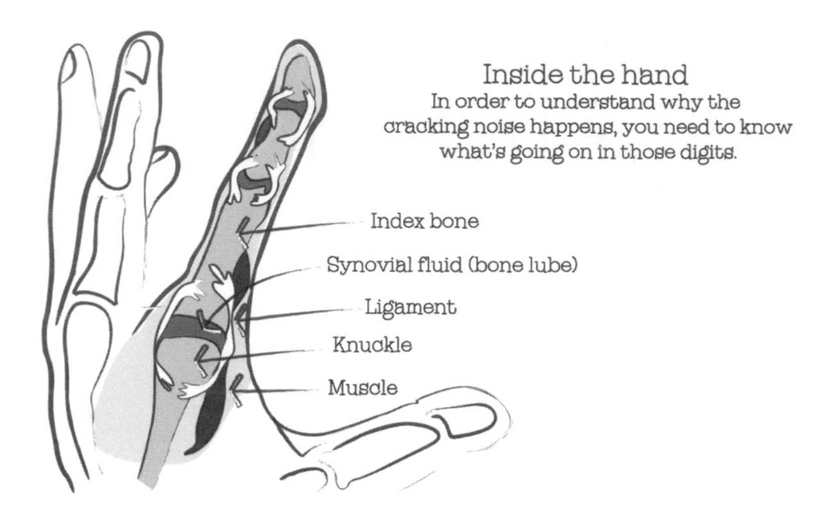

Inside the hand

In order to understand why the cracking noise happens, you need to know what's going on in those digits.

The three steps to cracking good success

1 Bones don't have hinges. They are held together by ligaments.

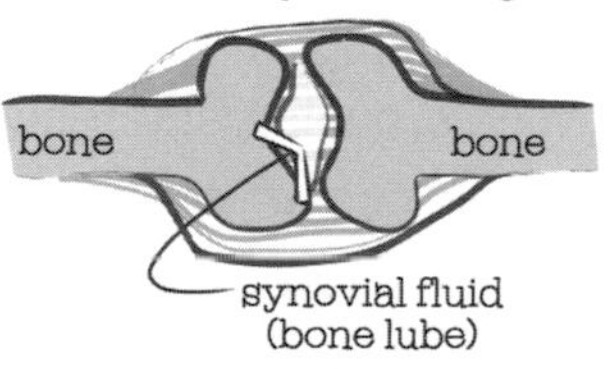

2 Inside each joint, there is a drop of lube called synovial fluid.

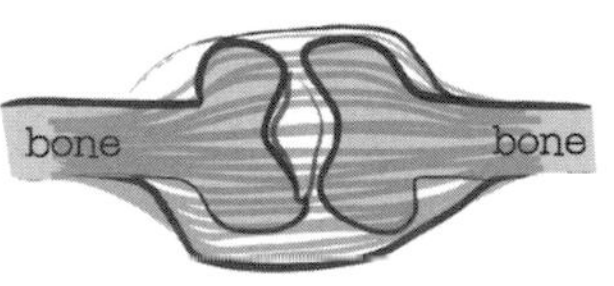

Synovial is held tight by the ligaments. It's kind of like fizzy drink trapped in a closed bottle.

3 When you pull finger bones apart, the ligaments stretch and the pressure inside the fluid reduces.

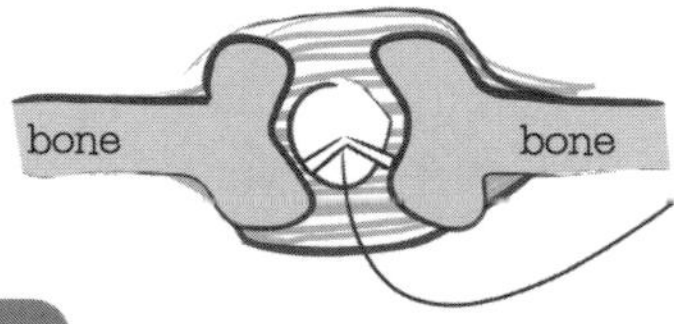

This makes room for a tiny bubble to form ... and this tiny bubble can make a BIG sound.

Cracking Knuckles

It has long been thought that cracking your knuckles leads to an increased chance of getting arthritis. This is far from being true. The real surprise is that those who crack their knuckles will end up with weaker hands (reduced grip strength).

the pressure drops, a bubble (mostly carbon dioxide) appears, in just a few thousandths of a second. As this gas bubble pops into existence, it will make a popping sound, which is the first of the two sounds.

The bubble takes up about 15% of the joint space which is now bigger. Because the joint space suddenly has a bubble in it, the liquid just as suddenly pushes on the ligaments — snapping them back outwards to their original position. This 'snapping back' of the ligaments gives the second sound.

The energy set loose inside the joint is fairly low — about 7% of what you need to damage the cartilage. But if you crack your knuckles repeatedly it might be a different story.

One study was carried out by Dr Daniel Unger, who had cracked the knuckles on his left hand for 50 years. He then compared his left hand (with the cracked knuckles) and his right hand (with no cracked knuckles). His left hand was no more arthritic than his right hand — but 'one person' is a very small sample size.

Another study looked at 300 people who had been cracking knuckle joints for 35 years. There were no extra cases of arthritis in the hands. They did have slightly swollen joints (which is no big deal). But the real surprise was that their hands were weaker — their grip strength was one-quarter as strong as it should have been!

So cracking your knuckles won't bother you in the short term, but 35 years from now, you might not be able to open a jar of Vegemite!

References

Brodeur, Raymond, 'The audible release associated with joint manipulation', *Journal of Manipulative and Physiological Therapies*, March/April, 1995, pp. 155–164.

Castellanos, Jorge & Axelrod, David, 'Effects of habitual knuckle cracking on hand function', *Annals of the Rheumatic Diseases*, vol. 49, 1990, pp. 308, 309.

Unger, Donald L., 'Does knuckle cracking lead to arthritis of the knuckles?', *Arthritis and Rheumatism*, May 1998, p. 949.

Curse of King Tut

From time to time, Hollywood releases another *Curse of the Pharaohs* or *Mummy* movie. They all have some kind of credibility, because of King Tut's well-known 'curse'. You see, when the tomb was opened in 1922, journalists reported that an inscription near the door of King Tutankhamen's tomb read: 'Death shall come on swift wings to him that touches the tomb of the Pharaoh.' The curse seemingly proved itself to be real, when all of the archaeologists and workers who desecrated the tomb of Tutankhamen were reported to have died horrible and early deaths.

About 3300 years ago, the tragic Boy King, King Tutankhamen, reigned very briefly in Egypt (1361–1352 BC), until he died at only 18 years of age. There were four 18th-Dynasty 'Amarna Kings', and he was the third. Because the 19th-Dynasty rulers didn't like the rulers of the 18th Dynasty, the Amarna Kings were publicly stricken from the list of the royalty. Monuments to King Tut were destroyed, and the location of his tomb was forgotten.

Its whereabouts had been well and truly forgotten by the 20th Dynasty. When the chief architect started cutting the rock for the tomb for Ramses VI, he unknowingly let the rubble tumble over King Tut's tomb.

King Tut's tomb was also forgotten, because he had been a very unimpressive ruler. This had an unexpected advantage 3300 years later — his tomb had been perfectly hidden from robbers and his considerable treasure remained untouched.

By November 1922, the archaeologist Howard Carter had spent seven frustrating years looking for King Tut's tomb in Luxor's Valley of the Kings. Eventually, his workers dug down four metres beneath the tomb of Ramses VI, where they found an entrance in the rock that led to a large passageway three metres high by two metres wide. They cleaned out the rubble, and at the twelfth step, they found the top of a sealed stone doorway.

This was exciting news and Howard Carter immediately invited his financier, Lord Carnarvon, to be at the site for the opening of the tomb. Carter and Carnarvon were both present on the evening of 24 November, when all the rubble was removed to reveal the stone door displaying the seal of King Tutankhamen. Once this door was opened, it took a further two days of hard work to clear rubble from another descending stairway. This time they found a second door, which had the seals of both the Royal Necropolis and Tutankhamen. The workers made a hole through the stone door and, using the

The Curse of King Tut

*Actually, newspapers invented the inscription near the door of the tomb that mentioned 'Death on swift wings', claiming that it was the 'Curse of the Pharaohs' that had killed him.

A noted archaeologist

Tut's Curse

There was no curse on King Tut's tomb. It was a media beat-up aimed to sell newspapers. Much the same way the tabloid press works today for that matter. Nice to see some things never change.

light from a candle, Carter peered in. Lord Carnarvon asked, 'Can you see anything?' Carter replied, 'Wonderful things.'

There was magnificent treasure in the anteroom. There was even more in the inner room, which took them another three months to enter. Lord Carnarvon himself opened this inner door on 17 February 1923. King Tut's mummified remains were inside three coffins. The outer two coffins were made of hammered gold fitted to wooden frames, while the innermost coffin was made of solid gold.

Lord Carnarvon died on 6 April 1923 from pneumonia, a complication from an infected mosquito bite. It was then that the newspapers invented the inscription near the door of the tomb about 'death on swift wings', and claimed that the 'Curse of the Pharaohs' had killed Lord Carnarvon. But there was no curse on King Tut's tomb. (Mind you, other tombs did have curses inscribed on them. A typical curse was: 'If anybody touches my tomb, he will be eaten by a lion, crocodile and hippopotamus.')

King Tut's treasures were exhibited in various museums around the world. When Arthur C. Mace of the Metropolitan Museum of Art in New York and George Benedite of the Louvre in Paris both died after showing the treasures of King Tut in their museums, the 'curse' was again blamed. The Curse of the Pharaohs was then blamed for the deaths of people who were only remotely connected with the expedition — such as Carter's previous secretary, Robert Bethnell, as well as Bethnell's father. The fact that Bethnell's father died at the ripe old age of 78 was ignored.

Since then, people have examined the Curse of the Pharaohs from a scientific point of view.

The famous sceptic and magician, James Randi, in his *Encyclopaedia of Claims, Frauds and Hoaxes of the Occult and Supernatural*, gives the names of all Europeans who were present when the tomb was opened and when they died. You might have heard of special statistical tables called 'actuarial tables'. They give your life expectancy, based on where you live, whether you smoke, how long your parents and grandparents lived and so on. Randi looked up the relevant actuarial tables for all the people who were associated with King Tut's tomb, and who then later died.

In fact, the people who were present at the opening of the tomb lived a year longer than the actuarial tables predicted. Howard Carter died at the reasonable age (for those days) of 66. Dr Douglas Derry who actually dissected the mummy died aged over 80. And Alfred Lucas, the chemist who analysed tissues from the mummy died at 79.

Other studies have shown no obvious effects on life expectancy of the people involved with this excavation.

Science could finally bury the Curse of the Pharaoh. Sadly, the only person to die an unreasonably early death was the Boy King, Tutankhamen.

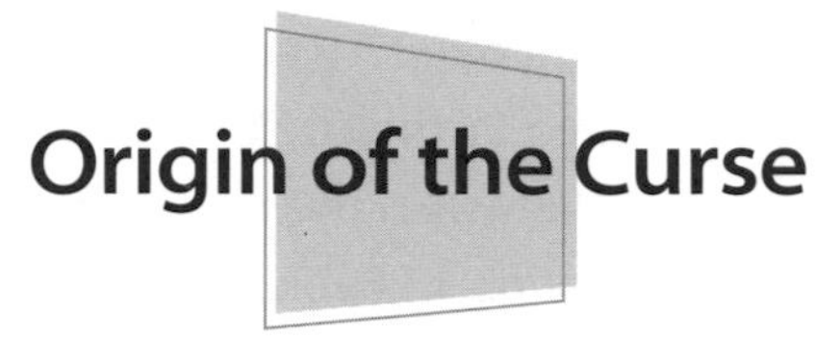

Origin of the Curse

The Curse was popularised by Hollywood movies, but it seems to have originated in books of fiction.

One possibility is the short story called *Lost in a Pyramid: The Mummy's Curse* written by Louisa May Alcott in 1860. She also wrote the novel *Little Women.*

Another possibility is a story by the American painter, Joseph Smith (1863–1950). He told of a curse on Tutankhamen's father-in-law, King Akhenaton. The throne passed from King Akhenaton after his death to his third daughter (the older two daughters had died). When Tutankhamen married the third daughter, the throne passed to him. King Akhenaton had deeply offended the priests, because he had interfered with their religion. He had combined the previous hundreds of gods into one single god, Ra, the disc of the Sun.

After Akhenaton died (and Tut had ascended the throne) the priests got their revenge by damning 'his body and soul ... to wander separately in space and never to be reunited for all eternity'. But this curse was aimed at Akhenaton, not Tut. The chief priest, Ay, took the throne when King Tut died — and there is speculation that he had King Tut murdered.

How to Measure a Curse

Dr Mark Nelson from the Department of Epidemiology and Preventative Medicine at Monash University wrote a paper on the curse of King Tut called 'One Foot in the Past. The mummy's curse: historical cohort study'. He has a doctorate in clinical epidemiology (statistics that study the health of human beings) and is very interested in archaeology and Egyptology.

He looked at the 44 Europeans identified by Howard Carter as being in Egypt on the relevant dates. Twenty-five of them were potentially able to be affected by the curse, because they were present at four significant events: '... the breaking of the seals and the opening of the third door on 17 February 1923, the opening of the sarcophagus on 3 February 1926, the opening of the coffins on 10 October 1926, and the examination of the mummy on 11 November 1926.' The people who were present at these events were deemed to be 'exposed to the curse'.

He then compared the life span of the exposed and non-exposed Europeans. There was no significant difference.

There may be a Curse of the Pharaohs or a Mummy's Curse, but it didn't make an appearance for King Tut.

References

'Science out to bury curse of pharaohs', *Sydney Morning Herald*, 15 September 2003, p. 8.

Marzuola, Carol, 'Old legend dies hard', *Science News*, vol. 163, 18 January 2003, p. 45.

Nelson, Mark R., 'The mummy's curse: historical cohort study', *British Medical Journal*, vol. 325, 21–28 December 2002, pp. 1482–1484.

Zombies

Zombies appear regularly on our TV screens, as more 'Tales of the Undead' are unleashed upon us. But here's a surprise — zombies can be real.

Zombies actually come from the Caribbean island of Haiti. They are people who have been almost-killed, and then later raised from the almost-dead by a voodoo priest, to be used as slave labour for the rest of their miserable lives. Zombies can move, eat, hear and speak, but they have no memory and no insight into their condition. Legends about zombies have been around for centuries, but it was only in 1980 that a legitimate case was documented.

The story begins in 1962 in Haiti. A man called Clairvius Narcisse was sold to a zombie master by his brothers, because Clairvius refused to sell his share of the family land. Soon afterwards Clairvius 'officially' died and was buried. However, he had then been 'unburied' and was actually working as a zombie slave on a sugar plantation with many other zombies. In 1964, after his zombie master died, he wandered across the island in a psychotic daze for the next 16 years. The drugs that made him psychotic were gradually wearing off. In 1980, he accidentally stumbled across his long-lost sister in a market place, and recognised her. She didn't recognise him, but he identified himself to her by telling her early childhood experiences that only he could possibly know.

Dr Wade Davis, an ethnobiologist from Harvard University, went to Haiti to research the story and discovered the process of making zombies. First, make them 'dead' and then make them 'mad', so that their minds are malleable. Often, a local 'witch doctor' secretly gives them the drugs to achieve this.

He makes the victims 'dead' with a mixture of toad skin and puffer fish. This can be added to the food, or rubbed onto the skin, especially the soft, undamaged skin on the inside of the arm near the elbow. The victims soon appear dead, with an incredibly slow breath, and an incredibly slow and faint heartbeat. In Haiti, people are buried very soon after death, because the heat and the lack of refrigeration make bodies decay very rapidly. This suits the zombie-making process. You have to dig up the bodies within eight hours of the burial, or they will die of asphyxiation.

You make the victims mad, by force-feeding them a paste made from datura (jimsons weed). Because datura breaks your links with reality, and then destroys all recent memories you don't know

Make your own Zombie at home

Here's one we prepared earlier

Project ingredients: One 'willing' participant, a mixture of toad skin and puffer fish (to be added to food or rubbed on skin). Once appearing as dead (don't use too much), you then make the victim insane by force feeding him a paste made from datura.

Zombies

Zombies come from the island Haiti. They are folk who have been almost-killed and then later raised from the almost-dead by a voodoo priest, to be used as slave labour for the rest of their miserable lives.

what day it is, where you are and, worst of all, you don't even know who you are. The zombies, now in a state of semi-permanent induced psychotic delirium, are sold to sugar plantations as slave labour. They are given datura again if they seem to be recovering their senses.

It all seems a little like the 'soma' drug given in Aldous Huxley's book, *Brave New World* — but without the happiness.

The Chemistry of Zombification

The skin of the common toad *Bufo bufo bufo* can kill — especially if the toad has been threatened. There are three main nasties in toad venom — biogenic amines, bufogenine and bufotoxins. One of their many effects is that of a painkiller — far stronger than cocaine. A story in Boccaccio's *Decameron* tells of two lovers who die after eating the herb, sage, that a toad had breathed upon.

The puffer fish is known in Japan as *fugo*. Its poison is called 'tetrodotoxin', a deadly neurotoxin — its painkilling effects are 160 000 times stronger than cocaine. Eating the fish can give you a gentle physical 'tingle' from the tetrodotoxin — and in Japan, the chefs who prepare *fugo* have to be licensed by the government. Even so, there are rare cases of near deaths or actual deaths from eating it. The toxin drops your body temperature and blood temperature, putting you into a deep coma. In Japan, some of the victims have recovered a few days after having been declared dead.

Datura (jimsons weed, Angel's Trumpet, *Brugmansia candida*) contains the chemicals atropine, hyoscyamine and scopolamine, which can act as powerful hallucinogens when administered in the appropriate doses. They can also cause memory loss, paralysis and death.

The person who administers these chemicals has to be quite skilled, so that they don't accidentally kill the victim. There is a very small gap between appearing to be dead and actually being dead.

References

Caulfield, Catherine, 'The chemistry of the living dead', *New Scientist*, 15 December 1983, p. 796.

Isbister, Geoffrey K. et al., 'Puffer fish poisoning: a potentially life-threatening condition', *Medical Journal of Australia*, 2/16 December 2002, pp. 650–653.

Morgan, Adrian, 'Who put the toad in toadstool', *New Scientist*, 25 December 1986/1 January 1987, pp. 44–47.

Wallis, Claudia, 'Zombies: do they exist?', *Time*, 17 October 1983, p. 36.

Dissing the Dishwasher

Water is an absolutely critical resource — we cannot live without it. Whenever there's a drought, and water restrictions are applied, people decide to conserve water by not using their dishwasher. But in reality, they would actually save water with a modern dishwasher.

The machine dishwasher was invented by a sophisticated ex-Chicago socialite, Josephine Garis Cochrane, who was married to a small-time politician in the small prairie town of Shelbyville, in Illinois. Unfortunately, her servants were constantly breaking her expensive heirloom crockery.

She was able to do something about it, because creative engineering ran in her veins. Her grandfather (John Fitch) was an early steamboat inventor, while her father (John Garis) was a civil engineer who helped build Chicago. She designed a machine that squirted jets of hot soapy water onto crockery, held in place in a wire rack. Her dishwashing machine worked well, and won the highest award at the 1893 World Exposition for 'the best mechanical construction, durability, and adaptation to its line of work'.

At first, only large hotels and restaurants used her machines from the Garis-Cochrane Dish-Washing Machine Company. (Her company later changed its name to Kitchen Aid, and was eventually taken over by Whirlpool.)

The really big market — households — initially resisted the dishwasher for both technological and social reasons. First, in the early 20th century, most houses didn't have enough hot water on tap to run a dishwasher. Second, the water was 'hard' in many parts of the United States, and wouldn't easily make suds. But third, and most importantly, the women of the early 1900s actually enjoyed washing dishes with others at the end of a meal, using it as a mini social occasion.

Dishwasher machines entered the home market only in the 1950s. By then, the advertising pitch had changed to emphasise that the machine could kill germs, because it used water that was much hotter than could be easily tolerated with hand dishwashing. In addition, the prosperity in the United States following World War II made leisure time and independence more important to the American housewife. By 1969 (the year of the Boeing 747, the first Moon landing and the Concorde), dishwashers were regularly being installed in new dwellings.

The early dishwashing machines did indeed use lots of water — 70 litres or more. But, according to recent test reports in *Choice* magazine, modern dishwashers use between 16 and 24 litres on a

You ... not so dirty ... rat

The modern dishwasher

Dishwashers

During times of drought, when water restrictions are applied, people decide to conserve water by not using the dishwasher. But in reality, they would actually save water by using a modern dishwasher.

'normal' cycle — and can use even less on the economy cycle. If you accumulate dishes from each meal until you have a full load, your dishwasher will use less water than if you had washed up by hand after each meal — and for around 20 cents per day. Of course, you shouldn't rinse the plates with water — merely dry scrape them — before you put them in the dishwasher.

Drying the dishes with heat at the end of the washing cycle uses a lot of energy. You can use less energy if you select the 'air-dry' option from the control panel — or you can simply open the door at the end of the washing cycle and let the plates dry slowly in the air. You can use even less energy if you choose the lowest possible temperature — but you have to use a rinse aid to stop spots or a film forming on the plates. And of course, it's better to use the dishwasher late at night when the electrical demand on the grid is low.

Even so, I miss the quiet relaxing companionship that comes with a team of two people washing and drying the dishes.

Man vs Machine

Rainer Stamminger, Professor of Home Economics at the University of Bonn in Germany, recently carried out the definitive test that proved the superiority of the mechanical dishwasher over the human dishwasher. He set up the typical daily washing load of a family of four — some 140 items including pots, pans, plates, glasses and cutlery, lovingly splattered with dried residue of egg, spinach, oats, etc. Then, 75 volunteers from nine countries washed them by hand — with surprising results.

First, the water used ranged from 15–345 litres — more than the modern European dishwasher on the economy cycle (12–20 litres).

Second, the machines used less than half the electricity (about 1 kWh) of the hand-washers (about 2.4 kWh).

Finally, only 15% of the hand-washers could reach the cleanliness of the machine dishwasher.

There were national differences. The British were the quickest, while the Turks were the slowest (at 108 minutes). The Spanish (followed closely by the Turks) were the cleanest. Surprisingly, the Germans (with their image of obsessiveness) achieved only mediocre cleanliness.

Efficient vs Non-efficient

There is one problem with the newer, more efficient dishwashers. The spray arms clog up.

The older, less efficient dishwashers simply dumped the used water. The newer machines use less water because they reuse the water after filtering it. But this means that the tiny holes in the spray arms that squirt out water sometimes clog up. Either very small particles of food join up to make bigger particles, or the bigger particles can sneak through the filter.

So when the cups and the plates start coming out dirty, check the instruction manual for how to clean the holes in the spray arms.

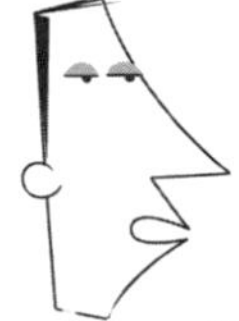

References

'Dishwashing made easy', *Choice*, June 2002, pp. 38–41.

Binney, Ruth, 'The origins of everyday things', *Readers Digest*, London, UK, 1998, p. 39.

Panati, Charles, *Panati's Extraordinary Origins of Everyday Things,* Harper & Row, New York, USA, 1987, pp. 103–104.

Aluminium and Alzheimer's Disease

Both aluminium and Alzheimer's disease begin with 'al' — about the only link between them. However, some people claim that the link is much stronger, and that aluminium causes Alzheimer's disease.

Aluminium is (after oxygen and silicon) the third most common element (at 8%) in the Earth's crust. As a result, you cannot avoid being exposed to it. You will find aluminium in drinking water, foods, pharmaceuticals, antiperspirants and printing inks. It is used to dye fabrics, preserve wood, distil petroleum, and to make rubber, paint, explosives and glass.

Aluminium has had bad press for a long time, especially since the 1920s. Rudolph Valentino's death in 1926 at the tender age of 31 was blamed on aluminium poisoning from aluminium cookware — but he actually died from a perforated stomach ulcer. Howard J. Force, a self-proclaimed 'chemist', added momentum to the anti-aluminium movement with pamphlets such as 'Poisons Formed by Aluminum Cooking Utensils'. It was probably not a coincidence that he also sold stainless-steel cookware.

The first scientific 'evidence' about the toxicity of aluminium appeared in the mid-1970s. At autopsy, the brains of people suffering from Alzheimer's disease were studied and found to have high concentrations of aluminium — and almost always in characteristic neuro-fibrillary tangles in the nerves. Did the aluminium cause Alzheimer's disease? No. It eventually turned out that the neuro-fibrillary tangles were very 'sticky' — and absorbed the aluminium from the water used to wash them during the autopsy pathology process.

Around the same time, a brand-new aluminium-related disease appeared — dialysis encephalopathy. Patients with chronic kidney failure were now routinely being treated with a new technique called dialysis. It used hundreds of litres of water each day to purify the blood. Unfortunately, the aluminium naturally present in the water entered the blood, and could not be removed — because the kidneys weren't working. As the blood levels of aluminium soared to thousands of times higher than normal, the patients became confused and demented. As soon as the problem was recognised, the condition dialysis encephalopathy was prevented by removing the aluminium from the water.

Aluminium and Alzheimer's Disease

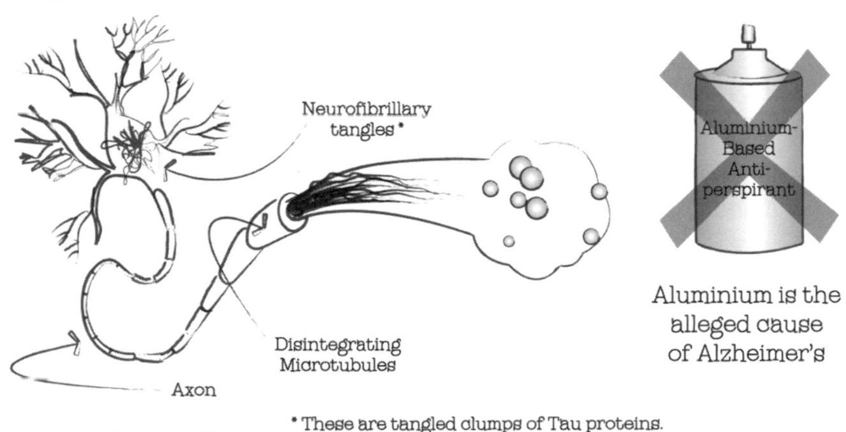

* These are tangled clumps of Tau proteins. In healthy cells, these tangles do not exist.

Alzheimer's

Both Aluminium and Alzheimer's begin with 'Al', that's about the only link between them. Aluminium has had bad press for a long time about how it can cause Alzheimer's disease . . . which is simply not true.

Administering aluminium in massive concentrations directly into the blood of very sick people with failed kidneys did cause dementia. But there are many causes of dementia, including Alzheimer's disease. Sufferers of Alzheimer's disease have typical changes in the brain that can be seen only under a microscope. These changes are called 'neuro-fibrillary tangles'. Although the dialysis patients, had very high aluminium levels and dementia, they did not develop the neuro-fibrillary tangles.

The average intake of aluminium is about 10–50 mg per day. But even people who take antacids and buffered aspirin, which elevate their aluminium intake to 1000 mg per day, have no increased incidence of Alzheimer's disease.

Dr Charles DeCarli, the director of the Alzheimer's Disease Center at the University of Kansas Medical Center says, 'In my opinion, the supposed relation between aluminum and Alzheimer's disease is a simple case of neuromythology.'

Congealed Electricity

Although aluminium is very cheap nowadays, it was once very precious.

Aluminium was first isolated in 1825 by Hans Christian Orsted. In 1855, it was shown to the general public at the Paris Exposition. The aluminium was separated from its mineral with chemistry and huge amounts of electricity. As electricity became cheaper, so did aluminium. By the 1960s, aluminium became the most produced metal (after iron) in the world. It has been nicknamed 'congealed electricity', because of the enormous quantities of electricity needed to make it.

In the mid-1800s, aluminium was used for its rarity value and resistance to corrosion. On 4 July 1848, the cornerstone was set for the Washington Monument in Washington, DC. This towering

169-metre tall obelisk is the tallest freestanding stone structure in the world. When it was finally dedicated on 21 February 1885 (after financial setbacks, including the Civil War), it was capped with a tiny 23-centimetre pyramid of aluminium — because aluminium was so precious.

References

Peder Flaten, Trond, 'Aluminium as a risk factor in Alzheimer's disease, with emphasis on drinking water', *Brain Research Bulletin*, 2001, vol. 55, no. 2, pp. 187–196.

Soni, Madhusudan G. *et. al.*, 'Safety evaluation of dietary aluminium', *Regulatory Toxicology and Pharmacology*, 2001, vol. 33, pp. 66–69.

Encyclopaedia Britannica, (DVD), © 2004.

Bermuda Triangle

When compared to the rest of the world, the Bermuda Triangle is quite large, but its reputation is enormous — much bigger than its size would indicate. Geographically speaking, the triangle is an area in the Atlantic Ocean bounded by Bermuda, Puerto Rico and Miami in Florida. However, the myth of the mysterious disappearances associated with the Bermuda Triangle is so powerful that books, TV documentaries and even movies have been made about it.

The seeds of the myth began at 2.10 pm, on 5 December 1945, when a flight of five Avenger Torpedo bombers lifted off from the airstrip of the naval base at Fort Lauderdale in Florida, on a routine bombing-training run. The story goes that in perfectly clear weather, these experienced aviators became mysteriously disorientated, and in a series of increasingly panicky radio transmissions, asked for help. The last radio transmission from Flight 19 was at 7.04 pm. By 7.20 pm a Martin Mariner rescue plane was dispatched — and it too vanished without a trace. (By the way, the missing pilots and their missing planes made a brief appearance in the movie, *Close Encounters of the Third Kind*, where it was implied that they had been abducted by aliens.)

But the myth claims that not only planes vanish in the area. Many ships apparently came to foul ends in the Bermuda Triangle, including the 19th-century sailing ship, the *Marie Celeste*, which was supposedly found drifting and abandoned in perfect condition.

The Bermuda Triangle has moved with the times, and since then, many more ships, including the nuclear submarine USS *Scorpion*, have vanished there without a trace.

The real story is more prosaic.

First, the Bermuda Triangle is huge — over one million square kilometres, or one-fifth the area of Australia (or the contiguous continental United States). You can fit a lot of ships in an area that size!

Second, the triangle is just north of the birthplace of most of the Atlantic hurricanes that lash the east coast of the United States. The Gulf Stream, the 'river in a sea', flows swiftly and turbulently through the Bermuda Triangle, dumping huge amounts of energy there. Many wild storms can suddenly burst into existence, and can, just as suddenly, fade away.

Third, the undersea landscape is incredibly varied, ranging from shallow continental shelf to the deepest depths of the Atlantic

They were there ... just a minute ago!

Lt Charles Taylor
(With a hangover)

Bermuda Triangle

Flight 19, 5 December 1945 ... a collection of non-experienced trainees head out on a training mission led by a hungover commander ... not the smartest move.

Ocean, about 30 000 ft (9144 m) deep. This means that some wrecks would be very difficult to find.

Fourth, it's one of the heaviest-travelled routes for pleasure craft in the world. So you would expect many nautical mishaps.

Fifth, a survey by insurance underwriters Lloyds of London shows that, on a percentage basis, there are no more ships lost in the Bermuda Triangle, than anywhere else in the world.

When you examine the specific stories more closely, the myth unravels even more.

The *Marie Celeste* was found abandoned on the other side of the Atlantic Ocean, between Portugal and the Azores. Contrary to legend, its sails were in very poor condition, and it was listing badly — definitely not in near perfect condition. The USS *Scorpion* was found, sunk, near the Azores, again, a long way from the Bermuda Triangle.

The story of Flight 19 on 5 December 1945, is the key.

The naval aviators were not experienced. They were all trainees, apart from the commander, Lt Charles Taylor. Reports say that he was suffering from a hangover, and tried unsuccessfully to get another commander to fly this mission for him. The weather was not clear — a sudden storm had raised 15-m waves. The Avenger Torpedo bombers simply got lost, ran out of fuel and sank in the storm, after dark, and in high seas. One of Commander Taylor's colleagues wrote, '... they didn't call those planes "Iron Birds" for nothing. They weighed 14 000 pounds (over 6 tonnes) empty. So when they ditched, they went down pretty fast.'

The Martin Mariner rescue plane sent to look for the Avengers did not vanish without trace. These rescue planes were virtually flying fuel tanks, because they had to remain aloft for 24 hours continuously. And prior to this incident, they had a reputation for leaking petrol fumes inside the cabin. The crew of the SS *Gaines Mill* actually saw the Mariner breaking up in an explosion about 23 seconds after take-off, and saw debris floating in the stormy seas. After this Mariner plane exploded, the Navy grounded the entire fleet of Martin Mariners.

The myth of the malevolent supernatural powers hiding in the Bermuda Triangle began when Vincent H. Gaddis wrote rather creatively about Flight 19 in the February 1964 issue of *Argosy: Magazine of Masterpiece Fiction* in a story called, 'The Spreading Mystery of the Bermuda Triangle'. But the myth really took off in 1974, when Charles Berlitz released his bestseller *The Bermuda Triangle*, an even more imaginative account.

Exotic explanations for these disasters include power crystals from Atlantis, hostile aliens hiding under the waters, violent vortices from other dimensions, and evil human beings using anti-gravity machines.

At least it's true that the stories are far more interesting than the real explanation, but that's about the only truth to them.

Ice That Burns

The Bermuda Triangle doesn't sink ships. But there is something very strange lurking under the ocean floor of the Bermuda Triangle — an ice that burns.

It's called a 'methane hydrate' — a single molecule of methane trapped in a cage of six water molecules. Methane has the chemical formula of CH_4 — one atom of carbon surrounded by four atoms of hydrogen — while water is your standard H_2O. If methane and water are together in the same place, and if the pressure is high enough and the temperature low enough, you can get a methane hydrate. If you bring a lump of methane hydrate to the surface, the icy water melts releasing the methane, which will burn quite nicely. In one sense, these methane hydrates are a little like vampires — they will fall to pieces if you bring them out into the light.

It was only in the late 1960s that Russian scientists discovered natural hydrates in the freezing Siberian permafrost. In the 1970s methane hydrates were discovered at the bottom of the Black Sea. The Black Sea is loaded with these methane hydrates. In fact, sailors have long reported seeing bolts of lightning set fire to the methane on the surface of the sea — methane that had spontaneously bubbled up from below the ocean floor. And since then, they have been found in many, many places under the ocean floor, including the notorious Bermuda Triangle.

These methane hydrates are now the largest untapped source of fossil fuels remaining on Earth.

References

Gaddis, Vincent, 'The Deadly Bermuda Triangle', *Argosy*, February 1964.

US Department of the Navy — Naval Historical Center:
www.history.navy.mil/faqs/faq8-1.htm
www.history.navy.mil/faqs/faq8-2.htm
www.history.navy.mil/faqs/faq8-3.htm

Prayer Heals

Many people believe that prayer to a Higher Being can heal. This belief goes back a long way. In the Bible, *Genesis* 20:17 says: 'Then Abraham *prayed* to God, and God healed Abimelech, his wife and his slave girls so they could have children again.' But how much faith can we place in this belief?

In 1872, the scientist F. Galton wrote a paper called 'Statistical inquiries into the efficacy of prayer'. He found no benefits, or disadvantages, from prayer. Since then there have been other studies by medical doctors, all claiming to show the benefits of prayer. For example, William S. Harris and his colleagues at the Mid America Heart Institute at St Luke's Hospital in Kansas City, wrote a paper in the *Archives of Internal Medicine*, claiming to show that prayer improved the outcomes of patients admitted to the hospital's coronary care unit.

However, a brief look at the paper shows major flaws in the science.

First, the patients were supposedly 'randomly' assigned to either the prayer group or the non-prayer group — but they weren't. The random assignment was based on whether the last digit of the medical record number was odd or even. Odd and even numbers are not random.

Second, the people who did the praying (intercessors) were not monitored to see that they really did pray. There are two problems here. First, you wouldn't do a drug trial by delivering a kilogram of drugs to the hospital store, and then hope that the hospital staff get the right drugs to the right patients, in the right doses, and at the right times. In addition, the intercessors were supposed to pray for a person by their first name, e.g. Fred, Linda, etc. In any group of 1000 people, there are many shared first names. How did the prayers get directed to the right Fred or Linda?

Third, the study also claimed that the pre-existing illnesses of the patients were roughly similar. However, they were not similar. The people who were prayed for were not very sick. However, the people who were not prayed for, were much sicker. They had a far greater incidence of previous heart attacks, unstable angina, pulmonary oedema, hypertension, cardiac arrest, diabetes, chronic renal failure and so on. This happens when selection is not random.

Fourth, a few of the patients who had been prayed for had extremely long hospital stays (up to 161 days). They did not get better quickly. So to make the statistics look better, these patients were dropped from the study!

Can you feel the power ? No ... aren't you paying ... I mean praying enough?

Prayer Heals

Many folk believe that prayer to a Higher Being can heal. To date, no studies have been able to prove this. It still remains that the scientific links between health, religion and spirituality are still inconsistent and weak.

Finally, some of the patients in the prayer group got better before the intercessors had a chance to pray for them! So either the prayer was giving results before it was started, or perhaps there was something wrong with the study.

Not everybody thinks of prayer in the same way. For example, the cynical Ambrose Bierce gave this definition in *The Devil's Dictionary!* 'Prayer? To ask the laws of the universe to be annulled on behalf of a single petitioner confessedly unworthy.'

Perhaps one day a study will show that prayer does have healing powers. So far, none of the studies have. But that would be something to pray for.

Prescription Prayer

At various times around the world, medicine and religion have been closely intermingled — and at other times and places, they have been kept apart. In the case of the United States, 95% of the population believes in God, and 80% pray regularly and believe that God does his work through medical doctors. In one poll of 1000 American adults, 790 believed that spiritual faith could help recovery from a disease.

Certainly, there can be a link between your degree of religiousness and your degree of health. For example, studies show that some groups in society (e.g. Mormon elders, Roman Catholic priests and Trappist monks) have better health and longer lives. This is quite predictable, because to enter one of these groups, you have to promise to follow their standards of conduct. Most of these standards (such as avoiding smoking and alcohol, and sometimes meat) are associated with lowering risks to your health.

Many studies show that you recover better from a cancer if you do yoga, or have an optimistic attitude. And for many people, religion can give them this optimistic state — which can only help.

However, in the early part of the 21st century, the scientific links between health, religion and spirituality are still inconsistent and weak.

References

Harris, William S., *et. al.*, 'A randomized, controlled trial of the effects of remote, intercessory prayer on outcomes in patients admitted to the coronary care unit', *Archives of Internal Medicine*, 15 October 1999, pp. 2273–2278.

Sloan, R.P., Bagiella, E. & Powell, T., 'Religion, spirituality and medicine', *The Lancet*, 20 February 1999, pp. 664–667.

Microwaves Cook From the Inside

Microwaves are very weird. They will make food hot, but they are not hot themselves! We started using fire to heat our food about one million years ago, and since then we have used a number of variations on this theme, e.g. baking, boiling, steaming, poaching, roasting, grilling and frying. There appeared to be no new way to cook food until we started using microwaves, about 50 years ago. Even today, most people don't really understand microwaves — perhaps because it's a 'new' process. Whatever the reason, most people mistakenly believe that microwaves cook the food from the inside.

The first real 'use' of microwaves was in British radar units during World War II. Radar gave the British forces the huge advantage of being able to 'see' approaching enemy planes at night, or through thick cloud.

Wartime radar began in 1940, when Sir John Randall and Dr H.A. Boot invented the magnetron, a device to generate microwaves. When the magnetron was incorporated into a functioning radar unit, it worked in short regular spurts. The radar unit squirted out the microwaves for only a brief instant, and then stopped transmitting. A different section of the radar unit then listened for an echo which occurred only if the transmitted radar beam happened to 'paint' (or land on) a target. Some of the radar beam energy would bounce off the target back to the radar unit. If the radar unit had to wait for the echo for a long period of time, the incoming planes were far away — but if there was only a short delay, then the planes were very close.

During World War II, the British government sought the assistance of the American government in the development of radar. The Raytheon Corporation in the United States became involved. Dr Percy L. Spencer, an engineer with Raytheon, redesigned the radar units, and worked out how to boost production from 17 units per week to 13 000!

The idea of using microwaves to cook food happened accidentally in about 1946. It was the same Dr Spencer who hit on the idea. He had been testing a magnetron and needed a break. Luckily, he had a stash of chocolate in his pocket — but not so luckily, the chocolate bar had melted, and had ruined his trousers. Why had the chocolate melted? After all, it wasn't a hot day.

Caution: Filling may be freakin' hot!

The Aussie pie
If heated in a microwave, a pie will have a really hot centre and be cool outside.

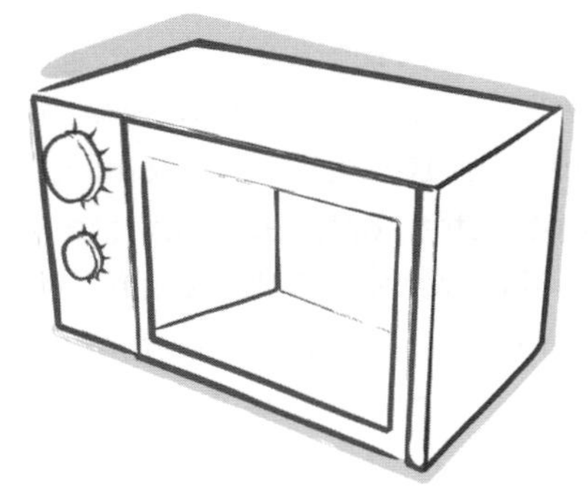

Microwave oven

Microwave

The mythconception that microwave ovens cook from the inside out is seemingly validated by the pie. Pastry is low in water, so when you heat one, the crust, or pastry won't get very hot, but the contents will.

He was an engineer with both an appetite and a good sense of curiosity. He took a bag of popcorn kernels, and blasted them with microwaves out of his magnetron. Soon, there was delicious popcorn all over the laboratory floor. He then tried cooking raw eggs in their shells, but the experiment wasn't so successful. The pressure inside the eggs rose so rapidly that they burst. The microwaves could cook food — with varying degrees of success.

Raytheon took up his ideas and developed a commercial microwave oven, the Radar Range. It was enormous (as big as a fridge, weighing 300 kg) but with a very small cooking volume (roughly the same as a modern microwave oven). The sales, unsurprisingly, were quite modest.

So how do microwaves cook?

The Raytheon engineers soon discovered that microwaves pass right through glass, paper, pastry, fats and most china. On the other hand, water absorbs microwaves very well indeed. The microwaves 'shake' the water molecules directly. The molecules of water vibrate about 2.45 billion times each second, and as they rub against each other, this friction produces the heat for cooking. This is how microwaves cook.

Do microwaves cook from the outside in?

Think of the food as being in spherical layers, like an onion. Let's assume that each layer is one centimetre thick, and that it absorbs 10% of the incoming microwave energy. After the first centimetre, only 90% of the energy is left. After the second centimetre only 81% is left, after the third centimetre only 73% and so on. You can see that most of the microwave energy is in the outside layers, with very little getting to the very centre.

Food in a microwave oven cooks from the outside to the inside. So how did this myth that microwaves cook from the inside start?

There are two possible explanations.

First, the egg that Spencer exploded did seem to cook from the inside. However this happened because it had a shell that was low in water, and an inner core that was high in water. The egg looked 'normal' until the water on the inside turned into steam, exploding the egg. In this case, the inside (the water) cooked, and the outside (the shell) did not.

Second, pastry and other fatty crusts are low in water. If you heat a baked potato or a meat pie in a microwave oven, the crust or pastry won't get very hot, but the contents inside will. As you bite into the potato or the pie, you pass through the cool (low water) crust or pastry into the hotter innards — and burn your tongue.

Microwaves are merely very short radio waves. Their wavelength is typically around 10 centimetres — approximately the distance across your knuckles.

RADAR stands for 'RAdio Detecting and Ranging'. The 'detecting' part means that it can find (or detect) objects. The 'ranging' part means that it can give you the distance to the object you have found. It can also tell you if the object is moving towards you, or away from you — and how fast.

Today, radar is used for air traffic control, and law enforcement on the roads. It helps planes and ships navigate, measures speed in industry and sport, and is used for tracking spacecraft, and even looking at planets.

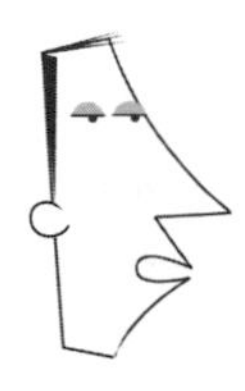

References

'Icy Sparks, the last word', *New Scientist*, no. 2119, 31 January 1988, p. 65.

Panati, Charles, *Extraordinary Origins of Everyday Things*, Harper & Row Publishers, New York, 1987, pp. 125–126.

Schizophrenia and Split Personality

In the movie, *Me, Myself and Irene*, Jim Carrey plays a state trooper with two personalities — Hank and Charlie. He is told that a split personality is part of his diagnosis of 'schizophrenia with involuntary narcissistic rage'. And indeed, according to a Harris Poll conducted for the National Organization on Disability in the United States, about two-thirds of people surveyed believe that split personality is part of schizophrenia. However, medically speaking, 'split personality' has nothing to do with 'schizophrenia'.

In the 19th century, Robert Louis Stevenson wrote about the two personalities of Dr Jekyll and Mr Hyde in the body of one person. The term 'split personality' re-entered the popular language in 1957 when C.H. Thigpen and H.M. Checkley wrote their famous book, *The Three Faces of Eve,* based on one of their patients. The popular concept was that the person could oscillate between two or more quite different personalities. Back then, the disorder was called Multiple Personality Disorder (MPD). But in the fourth edition of *The Diagnostic and Statistical Manual of Mental Disorders* (the 'Bible' of psychiatry), it has been renamed Dissociative Identity Disorder (DID).

Even so, the diagnosis is somewhat controversial. Some psychiatrists believe that DID is rare or non-existent, while others say that it is much more common than originally assumed. One of the very first recorded cases of MPD was that of Mary Reynolds way back in 1817. Between 1817 and 1944 when a major review of MPD was reported by Drs W. Taylor and M. Martin in the *Journal of Abnormal and Social Psychiatry*, a total of only 76 cases of MPD had been diagnosed. And indeed, in 1984, Thigpen and Checkley (the authors of *The Three Faces of Eve*) wrote that they now doubted the accuracy of a diagnosis of MPD. In other words, 'split personality' may not even exist as a medical condition.

'Schizophrenia', on the other hand, is tragically far more common — affecting about 1% of the population. It is a serious illness which manifests itself as major disturbances in a person's thoughts, emotions, behaviour and perceptions. It usually first appears in adolescence or early adulthood, although it can appear later. It usually happens earlier in men, and later in women.

There are two major groups of symptoms — positive–active and negative–passive.

A divorce from reality

Schizophrenia is a serious illness which manifests itself as major disturbances in a person's thoughts, emotions, behaviour and perceptions.

Schizophrenia

The so-called split personality or schizophrenia tragically affects about 1% of the population. It's a serious illness which doesn't lead to split personalities, but rather a disorientation of time, place and person.

There are many positive–active symptoms. They include:

- delusions (they may believe that thoughts are being inserted into their mind, that they are Elvis, or that they are being followed);
- hallucinations (they may see or smell things that are not there, or hear voices telling them what to do or commenting on their actions);
- disorganised thoughts (their speech jumps irrationally from one topic to another, or they may create new words); and
- disorganised behaviour (they may wear multiple layers of clothes, or shout inappropriately in public).

Delusions occur in 90% of schizophrenics, auditory hallucinations in 50%, with visual hallucinations trailing at 15%.

Negative–passive symptoms include:

- withdrawal;
- loss of motivation;
- loss of feeling;
- speaking much less frequently, and what they do say may be repetitious or vague; and
- flat presentation (unchanged facial expressions, reduced body language and decreased spontaneous movements).

The popular myth that 'split personality' is part of schizophrenia may have arisen because the word 'schizophrenia' comes from the Greek words meaning 'split mind'. But the word 'split' here does not refer to 'split personality', but rather to the fact that the person is 'split from reality'.

The word 'schizophrenia' is still used incorrectly in Hollywood movies, books, music and theatre reviews, and sports pages and when referring to conflicting choices (e.g. spiritual schizophrenia, or the schizophrenia of a secret agent).

There are enough myths about schizophrenia in its own right. We certainly do not need to confuse this distressing condition of schizophrenia, with the very different and incredibly uncommon condition of 'split personality'.

Other Myths of Schizophrenia

1. Schizophrenia is caused by bad parenting, and lack of moral fibre.
   We still don't fully understand the causes of schizophrenia. There is a genetic component, and there are some poorly understood environmental and biological triggers.
2. People with schizophrenia are unstable and violent, and may go 'wild' without any warning.
   The vast majority of schizophrenics are not a threat to others, just like the vast majority of non-schizophrenics.
3. You can never recover from schizophrenia.
   Many cases are treatable. Indeed, hope for a recovery is part of the treatment.
4. Schizophrenics have an intellectual disability.
   No, they are entirely different conditions. Schizophrenia certainly can cause problems with abstract thinking and concentration, but it does not lower your intelligence.
5. Schizophrenia is caused by witchcraft, or evil spirits, or demonic possession.
   No. Neither is schizophrenia God's punishment for the sins of a family, nor is it a result of unrequited love.

References

American Psychiatric Association, 'Dissociative disorders', *Diagnostic and Statistical Manual of Mental Disorders*, 4th edition, Washington, DC, USA, 1994.

Campbell, P. Michelle, 'The Diagnosis of multiple personality disorder: the debate among professionals', *Der Zeitgeist: The Student Journal of Psychology*, www.ac.wwu.edu/~n9140024/CampbellPM.html

Pendegrast, Mark, 'Possessed by demons', *New Scientist*, 4 October 2003, pp. 34–35.

Pyramid Building

If you sit in front of the TV set for long enough, you will eventually see yet another program about the mysteries of the pyramids. The later Egyptian pyramids are amazing — very large, and close to absolutely perfect in every dimension. Some of the TV programs claim that the pyramids had to be built by an advanced civilisation. After all, they claim, the Egyptians didn't even have the wheel.

Some of the later pyramids were stupendous. The Great Pyramid of Khufu at Giza was built to house the body of the Pharaoh Khufu, who reigned from about 2590–2567 BC. It was originally about 146 m tall, roughly the height of a 45-storey skyscraper. Its 2 300 000 stones sit on a base covering 5.3 hectares — each side measuring about 230 m long. Each stone is about one cubic metre in volume, and weighs a few tonnes. How could mere human beings build such an immense structure, especially if they didn't have the wheel? Surely, the story goes, only an advanced civilisation could possibly be the builders of these absolutely perfect pyramids?

However, as it turns out, there are plenty of answers from the 'Land of the Believable'.

First, in 1996, Stuart Kirkland Weir wrote an article about pyramid building from an energy viewpoint, in the *Cambridge Archaeological Journal*. It was a straightforward Time And Motion Study. He worked out how much energy a person could deliver in a day, and how much potential energy there was in the seven million-or-so tonnes of stones. By potential energy, he meant the extra energy that an object gets when you lift it above ground level. He worked out that, in terms of working days, the Great Pyramid needed about 10 million person-days, or 1250 people over some 8400 days or 23 years. If you allow for holidays, accidents, and problems associated with friction, a workforce about eight times greater (say 10 000 people, which was less than 1% of the population of Egypt at the time) working for a quarter of a century would have enough time to finish the job.

The Greek historian, Herodotus, wrote that the pyramid-building force comprised about 100 000 people. On one hand, he could be wrong because he wrote some 2000 years after the Egyptian pyramids were built. On the other hand, 100 000 people would make the job a lot easier. And 100 000 people would soak up about 10% of the population, thus reducing unemployment and social unrest.

The Pyramid Building Scheme

Pyramid Building

The later Egyptian pyramids are amazing ... very large, and close to absolutely perfect in every dimension – which is why claims surfaced that the pyramids had to be built by an advanced civilisation.

Second, the Great Pyramid is not perfect. The sides vary in length by some 18 cm. It is not dead level, but slopes up very slightly in the southeast corner.

Third, archaeologists have actually discovered the quarries from which the stones were taken as well as the remains of the ramps that carried the stones to the upper levels of the pyramids. A scene on the walls of the tomb of the 12th-Dynasty monarch, Djehutihotep (about 20 centuries BC) shows this process in some detail. It shows a gigantic five-metre, 60-tonne statue of Djehutihotep fastened to a wooden sledge. There are four rows of labourers, 172 in all, pulling hard on ropes tied to the leading edge of the sledge. At the front of the sledge, there is a man leaning out over the feet of the statue, pouring some kind of liquid under the sledge to lubricate its forward progress. And of course, there's the boss, sitting comfortably on the knees of the statue, probably calling out encouragement or orders to the workers below.

The combination of 60 tonnes and 172 labourers, works out to each person pulling about 330 kg. Modern reconstructions show that if you use a lubricant, it is easy to get a coefficient of friction of 0.1. This means that each man was exerting only about 33 kg — quite a believable figure.

Finally, archaeologists have just begun to uncover a city that supported the labourers who built the Great Pyramid. They have found streets, houses, graveyards, bakeries and all the infrastructure needed to support a floating population of 20 000 people.

The self-proclaimed psychic, Edgar Cayce, declared that the pyramids were built in 10 500 BC by an advanced civilisation, who then concealed their secrets in the currently undiscovered 'Hall of Records' under the front paws of the Sphinx, and then vanished.

He was partly correct. They were built by an advanced civilisation — the Egyptians of 4500 years ago.

The Workers

For many generations, archaeologists have focused their attention on the high nobility of Egypt — the kings and queens.

But Zahi Hawass (director of archaeology at Giza) and Mark Lehner (archaeologist at Harvard University) have studied the workers. The labourers were not slaves, but citizens who were proud to be part of a national project. Hawass discovered a workers' cemetery in 1990, while Lehner has begun to uncover their ancient city.

Thanks to tomb inscriptions and labourers' instructions, we now have something similar to a modern personnel flow chart. The workforce was broken into groups who were each responsible for a specific part of the pyramid construction, e.g. raising the chamber walls, placing the interior granite roof on the walls, etc. The groups consisted of many crews, each crew having four or five smaller units. Each unit had its own name, 'Great One', 'Green One' 'Friends of Khufu' or even 'Drunkards of Menkaure'.

One priest wrote of the workers who built his tomb: 'I paid them in beer and bread, and I made them make an oath that they were satisfied.'

References

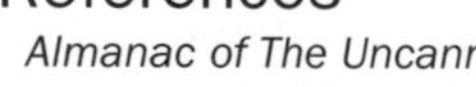

Almanac of The Uncanny, Reader's Digest Pty Ltd, Sydney, Australia, 1995, pp. 20–21.

'How the Great Pyramid was built', *How Was It Done?*, Reader's Digest Pty Ltd, 1995, pp. 318–324.

McClintock, Jack, 'Lost City', *Discover*, October 2001, pp. 40–47, 89.

Morell, Virgina, 'The Pyramid Builders', *National Geographic*, November 2001, pp. 78–99.

Astrology

The word 'disaster' comes from Latin and means 'malevolent star' — the Romans obviously thought that the stars could influence our daily lives. They were wrong, but even today, many people share this belief.

There are several types of astrology. 'General Astrology' looks at how humanity is affected by 'significant' alignments of the stars and planets. 'Genethlialogy' is a branch that looks at your life based on the positions of the stars and planets the moment you were born, while 'Catarchic Astrology' tries to find the most auspicious time to start a given task.

The idea that stars express 'divine will' originated some 2300 years ago with the Babylonians. With the naked eye they could see seven objects (that they called stars) moving through the sky — the Sun, the Moon, Mercury, Venus, Mars, Jupiter and Saturn. The Babylonians believed that the gods lived in these 'stars', and controlled the destinies of individuals and nations. They thought that the gods controlled us either directly by meddling in our affairs, or indirectly, by the intricate relationships of these stars with each other.

To describe the positions of these stars more easily, the Babylonians (who had a numbering system based on 12, not 10) divided the sky into 12 sectors. Today, we call these 12 sectors the 12 Houses of the Zodiac, e.g. Aquarius, Pisces, Aries and so on. The Babylonian astronomers/astrologers closely observed the sky decade after decade and noticed that these seven stars seemed to move through the Houses of the Zodiac in a totally predictable way — coming back to the same location in the same house at the same time every year.

Hence, the problem.

The constellations shift by about one degree every 72 years, thanks to the Earth's spin. The Earth spins around an imaginary spin axis that runs through the North and South Geographic Poles. But it doesn't spin true. If you have ever spun a top, you'll see that this spin axis soon begins to wobble. The spin axis will slowly sweep out a complete circle.

The same thing happens with the spin axis of the Earth — except that it takes about 26 000 years to sweep out a complete circle. So roughly every 2000-and-a-bit years (26 000 years divided by 12 Houses), the star signs get shifted by one House. The horoscopes you read in the daily newspapers (and often written by the most junior journalist on duty) are wrong by one House. You should be reading the previous star sign.

The famous French astrologer, Michael Gauquelin

Astrology

Michael offered free horoscopes to any reader of 'Ici Paris', if they would provide feedback on his accuracy. Thousands of identical horoscopes were sent out and 94% said they were very accurate. The horoscope sent was that of a mass murderer.

This is not a new discovery. In 129 BC, Hipparchus was the first to find this shifting-of-the-stars when he compared the astronomical records with what he saw with his eyes.

But most importantly, none of the detailed statistical studies that have looked at astrology have found any merit in it. For example, Bernard Silverman, a psychologist from Michigan State University in the United States, studied 2978 married couples and 478 couples who were divorced. He found absolutely no correlation between the divorce rates and those born under 'incompatible' zodiac signs.

The famous French astrologer, Michael Gauquelin, offered free horoscopes to any reader of *Ici Paris*, if they would provide feedback on the accuracy of his 'individual' analysis. He wanted to validate his profession scientifically. He sent out thousands of identical horoscopes, and 94% of the recipients replied that his reading was very accurate. What they didn't know was that the horoscope was that of a mass murderer, Dr Petiot, who admitted during his trial that he had killed 63 people.

While it is handy to blame misfortune on the stars, Julius Caesar was probably closer to the truth when Shakespeare had him say, 'The fault, dear Brutus, is not in our stars, but in ourselves ...'

Problems with Astrology

There are so many problems with astrology, that it's hard to know where to begin.

First, just when is a baby born? Is it when the mother's waters break, or when the baby's head appears, or when the baby's feet come out, or when the cord is cut, or when the placenta is delivered? For that matter, if the stars are indeed so powerful, why should the thin layers of the mother's abdominal wall and uterus provide any barrier to these supernatural forces?

Second, when reading the daily horoscope, how can one-twelfth of the world's population all fly down to Tasmania for a hot weekend with a tall handsome stranger? How would the airlines cope with the unexpected load?

Third, why are all the people born on the same day of the same year so different? Surely, they should all have similar appearance, lifestyle and behaviour?

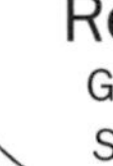

References

Gerrand, James, 'Correspondence on astrology', *The Skeptic*, vol. 5, no. 1, pp. 5–6.

Grey, William, 'Belief in astrology — a national survey', *The Skeptic*, Autumn 1992, pp. 27–28.

McKerracher, Phillip, 'Those who look to the stars have stars in their eyes', *The Skeptic*, vol. 3, no. 2, pp. 1–3.

Plummer, Mark, 'Reactions to astrology disclaimer', *The Skeptic*, vol. 5, no. 2, pp. 6–7.

Vince Ford, 'Astrology — the oldest con game (Part 1)', *The Skeptic*, vol. 5, no. 4, pp. 8–12.

Wheeler, Anthony G., 'Astrology and religion', *The Skeptic*, vol. 4, no. 2, pp. 3–4.

Williams, Barry, 'Planetary influences (astrology)', *The Skeptic*, Autumn 1992, pp. 12–16.

Use Your Brain

The human brain is one of the most complicated devices known. From a metabolic point of view it's a very expensive organ to run, requiring a lot of energy. Although it is only about 2% of our total body weight, it uses about 20% of our blood supply and 20% of our energy — and it generates about 20% of our heat.

There are many myths about this mysterious organ. One persistent myth is that we actually use only 10% of our brain — and that if we could use the remaining 90% we could each win a Nobel Prize or a gold medal at the Olympics, or even unleash our supposed psychic powers.

This myth has been circulating for nearly a century, and keeps re-emerging. Over the past decade, some 'motivational speakers' have shamelessly recycled this myth. They claim that if you enrol in their expensive course, you will suddenly be able to use all of your brainpower.

One of the earliest popular references to this myth is in Dale Carnegie's 1936 book, *How To Win Friends and Influence People*. He wanted to support his claim that if you worked your brain just a little harder, you could improve your life enormously. Without any neurological proof whatsoever, he boldly claimed that most people used only 15% of their brains. His book sold very well indeed, and helped promote the myth.

Dale Carnegie probably based his information on a misinterpretation of the experiments conducted by the neuropsychologist, Karl Lashley, in the 1920s. Lashley was trying to discover just where in the brain this strange thing called 'memory' was stored. He trained rats to run through mazes, and then measured how well they did as he removed more and more of the cortex of their brains. Lashley found that memory is not stored in one single place, but exists throughout the entire cortex, and probably a few other places as well. In fact, his results showed that the removal of any part of the cortex causes memory problems.

Karl Lashley's fairly straightforward results were somehow radically misinterpreted to show that the rats did well until they had only 10% of their brains left. There are two problems with this. First, he did not claim that we only need 10% of our brain. Second, he did not remove as much as 90% of any rat's brain.

Different versions of the myth are revived every decade or so. One version might have Albert Einstein, a certified mega-brain, saying (guess what?) that 'we only use 10% of our brain'. Or an

No Brain . . . No Pain!

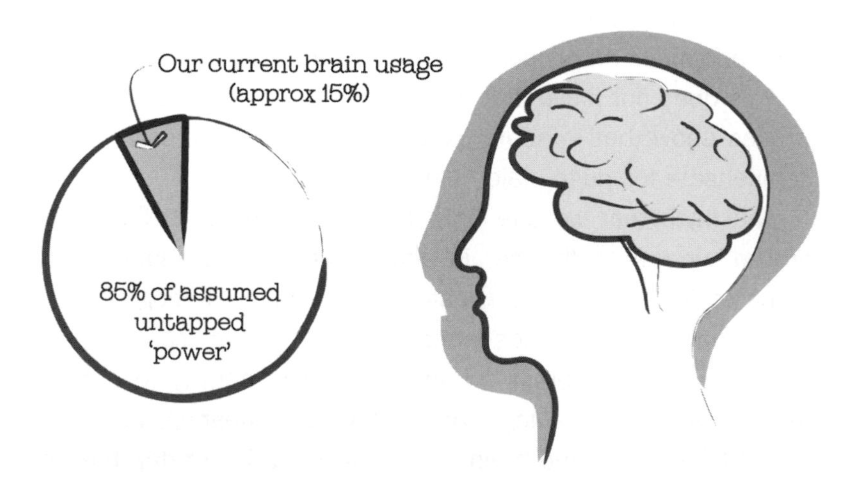

Use Your Brain

Dale Carnegie, the author of 'How To Win Friends and Influence People', claimed that if you worked your brain just a little harder, you could improve your life enormously. Without any proof, he boldly claimed that most folk used only 15% of their brain.

anonymous scientist (who is never named) has supposedly discovered that we use only 10% of our brain. Another version is that 10% of the mass of the brain is the conscious part of the brain, with the remaining 90% the subconscious part. (In reality, there is no such neat division.)

In the 1980s, Yorkshire Television in the UK showed a documentary called, *Is Your Brain Really Necessary?*. It described the work of the late British neurologist, Professor John Lorber, who specialised in children's illnesses. He saw many cases of hydrocephalus, a condition involving the constant circulation of cerebro-spinal fluid around the brain and spinal cord. If too much fluid is produced, or if its outlet from the brain is blocked, then it can build up inside the skull. This excess fluid usually makes the skull grow bigger. However, sometimes it just makes the brain meat get thinner as the meat gets squashed up against the bony skull.

Professor Lorber discussed many cases where young people did not have much brain, but had normal intelligence. In one extraordinary case, a young man had only a one-millimetre thickness of grey matter in his brain, instead of the average 45 millimetres. Even so, he had an IQ of 126 (the average is 100) and had gained an honours degree in mathematics!

This does not prove that most of the brain is useless. However, it does show that in some cases, the brain can recover from, or compensate for, quite major injuries.

The myth that we use only 10% of our brain is finally being proven untrue, with the invention of new scanning/imaging technologies (such as Positron Emission Tomography and Functional Magnetic Resonance Imaging) that can show the metabolism of the brain. In any one single activity (e.g. talking, reading, walking, laughing, eating, looking or hearing) we use only a small fraction of the brain — but over a 24-hour day, the entire brain will light up on the scan.

In fact, if you did use your entire brain at the one time, you would probably have a grand mal epileptic fit. And finally, you will never hear a doctor say, 'Luckily, he had a stroke in the 90% of the brain that is never used, so I think he'll be all right.'

Maybe 10% is wrong, and maybe it's right ...

The human brain has about a trillion neurons. These are the cells that we traditionally believe do all the 'thinking'.

These trillion neurons are 'supported' by 10 trillion other cells. They include the astroglia (which make up the scaffolding that holds the neurons in place, and which also feed the neurons with nutrition). Another class of support cell is the oligodendrocyte, which makes the insulating myelin sheath that wraps around part of the neuron. There are also epidydimal cells, that line the hollow chambers (called ventricles) in the brain. Other cells, the microglia, act as the brain's equivalent of the immune system.

The neurons make up 10% of the cells in the brain, while the supporting cells make up 90%.

It then gets more complicated. One strange finding from the autopsy examination of Einstein's brain was that he had more of these 'supporting' cells than average. Perhaps these supporting cells do a little thinking on their own ...?

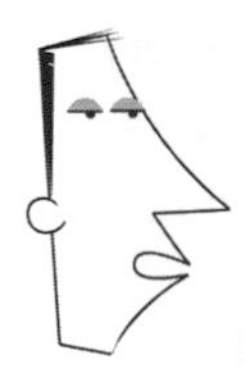

References

'Brain Drain', *New Scientist*, no. 2165, 19/26 December 1998–2 January 1999, pp. 85–86.

Radford, Benjamin, 'We use only ten percent of our brains', www.snopes.com/science/stats/10percnt.htm: Urban legends reference pages.

Sewell, R., Andrew, MD, 'Ten per cent', *Fortean Times*, March 2002, p. 54.

Plaskett, James, 'Is your brain really necessary?', *Fortean Times*, February 2002, p. 53.

Quantum Leap

If you move about in business circles, about once a month, you'll hear somebody say something like, '... and my plan will provide a huge quantum leap in performance'. The problem here is that a quantum leap is not huge — in fact, it's the smallest possible change.

Don't worry if you don't understand Quantum Mechanics — you are in very good company. That's because Quantum Mechanics Theory is very weird, often defying common sense. Consider the following statements: 'The electron that is flying around the centre of a hydrogen atom is indeed flying around the centre of that hydrogen atom — but at the same time, it is everywhere else in the Universe.' This statement is hard to believe, but true. You can see that Quantum Mechanics goes against the grain of every sensible atom in your body. But Quantum Mechanics works. Almost all electronic devices that we use today were developed with the insights gained from this theory. These include mobile phones, TVs, and the computers on your desk and in your car.

The concept of Quantum Mechanics was devised and evolved in the early 1900s by scientists such as Max Planck, Albert Einstein, Neils Bohr, Erwin Schroedinger and Werner Heisenberg. In fact, Michael Frayn's play, *Copenhagen*, is based on a conversation in 1941 between Bohr and Heisenberg about the possibility of building a nuclear weapon. This weapon could be built only because of the then-current knowledge of Quantum Mechanics. By the way, the version of Quantum Mechanics that most scientists adhere to today is the so-called 'Copenhagen Interpretation'.

Quantum Mechanics deals with the behaviour of matter and light on a very small scale — down deep in the land of atoms and subatomic particles. The times and distances involved are billions of billions of billions of billions times smaller than we deal with in our normal existence.

One idea fundamental to Quantum Mechanics is that energy can be exchanged only in little packets. You can have one packet, or two packets, but you cannot have one-and-a-half packets.

When you drive your car, it seems that you can vary the speed continuously. You can smoothly accelerate from 0 kph to 0.0001 kph, then to 0.0002 kph, and so on. But if you could scale your car down to the quantum scale of the very small, you would see that you can only vary the speed from one definite number (say 0 kph) to the next number above it (say 1 kph). There would be no speed between 0 kph and 1 kph, and you could not travel at

Quantum = Small

quan·tum:
The smallest amount of a physical quantity that can exist independently.

Quantum Leap

You may have heard someone say, 'This plan will provide a huge quantum leap in performance.' The problem with this saying is that a quantum leap is not huge. In fact, it is the smallest possible change.

any such intermediate speed. In the quantum world, your speed would suddenly jump from 0 kph to 1 kph.

This is almost certainly how the myth that quantum leaps are big came into existence. A quantum leap is a very definite leap. You have left one state of existence (say 0 kph) and re-appeared in a very different state (say 1 kph).

A quantum leap happens when a physical object moves from one quantum state to another quantum state. A quantum state is the unique condition of an object, based on a whole bunch of obscure and hard-to-understand physical quantities. Normally, we do not ever experience these quantities in the real world. These quantities include 'spin', 'charge', 'colour', 'strangeness' and 'charm'.

But remember, even though these quantities are incredibly obscure, and are often completely opposed to common sense, nevertheless, the computer at your desk or in your car can only be built if we have a small understanding of Quantum Mechanics.

So when business people talk about 'quantum leaps', they are really saying that there will be a definite change — and that it will be measurable only with the most sensitive devices made by human beings. So as the term 'quantum leap' went from the world of physics into common usage, it magically changed its meaning from 'very small' to 'very large'.

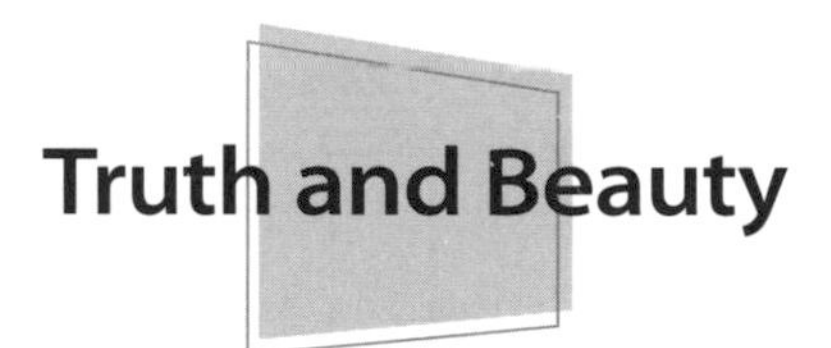

Truth and Beauty

Some years ago, I was invited to apply for a position with a Certain Company. The application process was going swimmingly. The Head of Certain Company even said that with my coming on board, things would go through a huge quantum leap. I ever so gently and politely pointed out that quantum leaps were not huge, but were very tiny. (After all, especially in science, we can't change the facts.)

His face turned black as his pupils simultaneously dilated. I was led to believe that I would not be employed by Certain Company.

In fact, I later heard that there would be 'blood on the walls' if I were employed. I guess that this is a good example of how 'the truth will set you free', because I was free of a job.

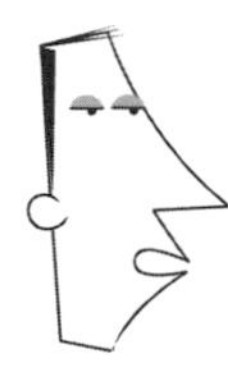

References

Encyclopaedia Britannica, (DVD), © 2004.

Gribbin, John & Chimsky, Mark, *Schrödinger's Kittens and the Search for Reality: Solving the Quantum Mysteries Tag*, Little, Brown and Company, Boston, USA, 1995.

White Spots on Nails

Most of us who have seen little white spots on our fingernails have been told that they are caused by very low levels of zinc in the body or not enough calcium. If you ask around, you'll get many more interesting opinions, including 'too much calcium', 'improper structural protein formation in the body', 'cuticles pushed back too hard', 'inactivity of the gland circulation' and 'dehydration of the nail from too much toluene or formaldehyde in the nail polish remover'. Others cite kidney disease, allergies, fungal infections, changing hormone levels, fasting, colds and viral infections. Even so, the most popular explanation is too little zinc and it's just as wrong as the other explanations.

Many animals protect the ends of their limbs with claws, talons or hoofs, which often come in handy as a weapon. The equivalent in human beings and other two-legged animals is the nail — a hard, horny plate on the back of each finger or toe. It protects the end of the finger or toe, helps you pick up small objects, and can be used as a scratching weapon.

Although nails appear to be simple, they are poorly understood, even today.

Nails are made mostly from a protein called keratin. On average, fingernails grow about one millimetre every 10 days. They grow faster in summer and slower in winter, faster in vigorous young adults and slower in the very young and the very old. All of the growth of the nail happens at its base, in the nail matrix.

Inside the nail matrix, new cells are constantly being generated and as they grow, they push the actual nail (called the nail plate) forward towards the end of the finger or toe.

The nail matrix is metabolically very active, and is sensitive to changes in your health. Your nails can reflect different aspects of your health by becoming thinner or thicker, cracked or furrowed. If you suffer major stresses or fevers, the nails may slow their growth dramatically, leaving horizontal grooves across your nails. These are called 'Beau's lines'. On the other hand, nail biters will be pleased to know that their bad habit actually speeds up nail growth. Quite a few heart and lung diseases can (for some unknown reason) change the shape of the nails so that the fingertips look like little clubs. This is called 'clubbing'. Indeed, you can make over a hundred different medical diagnoses by looking carefully at the nails.

So what about the white spots (officially called punctate leuconychia) on the nails? According to dermatologists, they have

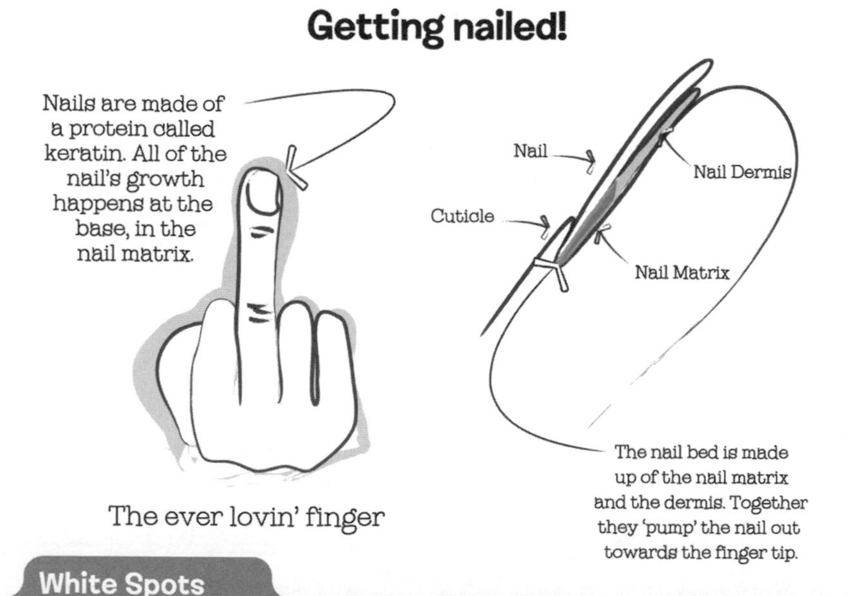

The ever lovin' finger

White Spots

White spots are mostly caused by minor damage to the nail matrix. The folk that try to sell you zinc supplements are wrong in nailing that particular diagnosis.

nothing to do with zinc levels. They are mostly caused by minor damage to the nail matrix, which makes the nail. The people who try to sell you zinc supplements are wrong in nailing that particular diagnosis.

A Few Nail Conditions

Furrowed horizontally: Beau's lines.

Furrowed vertically: trauma in the nail matrix leading to disruption or damage to the growing cells.

Thicker nails: whole nail thickened with sub-ungual (under the fingernail) hyperkeratosis (too much keratin), caused by, for example, psoriasis or a fungal infection. It's like the thickened skin 'scales' that happen in psoriasis.

Thinner nails: There can be several causes, including external forces (e.g. lamellar splitting of the nail or onycoschisis), diseases (e.g. lichen planus) or external factors such as nail products (hardeners) or false nails.

White lines and horizontal ridges: arsenic poisoning.

Blue nails: too much silver (e.g., from allergic rhinitis drops).

References

Baran, R. & Dawber, R.P.R., *Diseases of the Nails and their Management*, Blackwell Scientific Publications, 1994, pp. 75–76.

de Launey, W.E. & Land, W.A., *Principles and Practice of Dermatology*, Butterworths, Sydney, Australia, 1984, pp. 270–276.

Bible Code

Human beings love to see patterns — a dolphin in a cloud, the face of Jesus in a potato chip, and even a face on the surface of Mars. In his 1997 bestseller, *The Bible Code*, Michael Drosnin claimed that anyone could read the future in the Bible, by applying a simple mathematical code. The Bible he referred to was some of the books from the Hebrew Bible, such as *Genesis, Exodus, Leviticus, Numbers, Isaiah* and *Deuteronomy*. The code was amazingly simple. Start with any letter in *Genesis*, skip (say) three letters, write down the fourth letter, skip another three letters, write down the eighth letter, and so on. (Mathematicians call this a 'skip code'.) Keep going until recognisable words begin floating up eerily. If skipping three letters doesn't give you good words, try skipping four, or five, or any number you like.

Drosnin applied various lengths of skipping letters (his 'bible code') to the Torah. He claims that he found prophecies of the assassinations of President John Kennedy, Robert Kennedy and the Israeli Prime Minister Yitzhak Rabin, as well as prophecies about the Oklahoma City bombing, the rise of Hitler, and most of the significant events of the 20th century. Mind you, to extract the name RABIN, he had to skip 4771 letters at a time.

Drosnin was not the first to investigate this phenomenon. This 'bible code' stuff goes back a long way to the Kaballa — an esoteric Jewish mysticism of the 12th century. In 1994, Eliyahu Rips (an Israeli mathematician) reported in the journal, *Statistical*

Science, that if you used a skip code, you could find many recognisable words in *Genesis*. In 1995, Grant Jeffrey and Yacov Rambsel wrote *The Signature of God*, in which they applied a skip code of 20 letters to the book of Isaiah and found the phrase, 'Yeshua (Jesus) is my name'.

How could such a simple code predict the future so well?

First, this code has a high success rate in 'predicting' events after they have happened, and are already in the history books. In 1997, Drosnin predicted the end of the world in 2000 AD — or perhaps 2006, or perhaps later than 2006, or perhaps never. He covered every option with this broad prediction, guaranteeing that it will eventually come true. However, one of his many 'predictions' did come true — the assassination of Rabin.

Second, the Hebrew language, when written, usually omits short-sounding vowels, so you have to add them yourself. So Hebrew names can be written in many different ways, which increases your chances of success.

Halb Halb Halb played backwards is Blah Blah Blah

Jesus (or someone who looks like him) giving the inside scoop to the Bible Code

Bible Code

Stop Press: Apparently Bible Code is extremely accurate at predicting past events and occurrences.

Third, the code works on any book — not just the Torah. Professor Brendan McKay, a mathematician from the Australian National University, wrote about this in his paper, 'The Bible Code: Fact or Fraud'. McKay claimed that Drosnin's techniques are so broad that he could find 'prophetic messages' in any book. Drosnin, in a *Newsweek* interview, challenged: 'When my critics find a message about the assassination of a prime minister encrypted in *Moby Dick*, I'll believe them.' So McKay applied the skip code to *Moby Dick*. He found many famous assassinations in the novel, including that of Prime Minister Rabin, as well as John Kennedy, Martin Luther King Jr and Trotsky.

When McKay applied the skip code to Tolstoy's *War And Peace*, he found 59 words related to the Jewish *Hanukkah*. And Dr Rips, whose 1994 paper inspired Drosnin to write the Bible Code, has written: 'I do not support Mr Drosnin's work on the codes, or the conclusions he derives.'

If you want to try it yourself, go to Jeff Raskin's home page (www.jefraskin.com/forjef2/jefweb-compiled/published/bible_hoax.html) and plug in any text you like (the *Yellow Pages* or *White Pages* of the phone book, or just the latest bestseller) into his code program, and see how well you too can predict the past.

Lies and Statistics

This quote, often attributed to Mark Twain,'There are lies, damned lies, and statistics,' describes the bible code very well.

The original 1994 paper by Rips was published in a peer-reviewed journal, *Statistical Science*.'Peer-reviewed' usually means that the paper has to be reviewed and approved by a handful of anonymous experts in that particular field, before it can be published.

Although *Statistical Science* is peer-reviewed, it is a lively and interesting journal.The referees observed a curious result in Rips, paper — real names popping up at random, all too frequently — but could not immediately understand why. In the spirit of scientific exploration, they offered the paper to the readers as a 'challenging puzzle'. Indeed, it took several years before McKay and his colleagues found the flaws in the 1994 paper by Rips. They found many more flaws in Drosnin's book.

References

Fortean Times, August 1998, p. 7.

Allen, T.W., 'Bible Codes etc.', *The Skeptic*, Winter 2003, p. 65.

Bar-Hille, Maya, Bar-Natan, Dror & McKay, Brendan ,'The Torah codes: puzzle and solution', *Chance*, vol. 11, no. 2, 1998, pp. 13–19.

Williams, Barry ,'The Bible Code', *Australasian Science*, September 2003, p. 46.

Chocolate Zits

There's an old saying: 'You are what you eat.' But of course, eating carrots doesn't turn you into a carrot. The saying probably means that over-indulging in various foods can cause disease, e.g. coronary heart disease, diabetes, high blood pressure, dental caries and bowel cancer. But does chocolate give you acne?

Chocolate was first eaten 2600 years ago in northern Belize in Central America. We know this because, traces of chocolate have been found in several ancient ceramic vessels of the region. By the time of the Spanish Conquest in the 15th century, chocolate was consumed with most meals, usually mixed with other ingredients, such as maize or honey. Hernando Cortez, the Spanish conqueror of Mexico, brought back three chests of cacao beans to Spain in 1519. The Spanish were able to keep the manufacture of drinking chocolate a secret until 1606, when chocolate escaped from Spain and appeared in Italy. Then began its domination of the world. Today, chocolate is beloved by Antarctic explorers because it is rich in calories, and by most of us for its great taste and 'mouth feel'. It also contains caffeine, and other chemicals that directly affect the 'feel-good' neuro-transmitters in the brain.

But what about the almost-universal belief that chocolate is one of the most powerful producers of acne known?

The acne story begins in abnormal sebaceous glands, which happen to be much more active than normal sebaceous glands. A

Zits and The Zen of Chocolate

Chocolate was first eaten 2600 years ago in northern Belize in Central America.

A typical Belizean enjoying one of his many daily, chocolate fixes.

*My dark friend the chocolate cake.

The highly active sebaceous gland

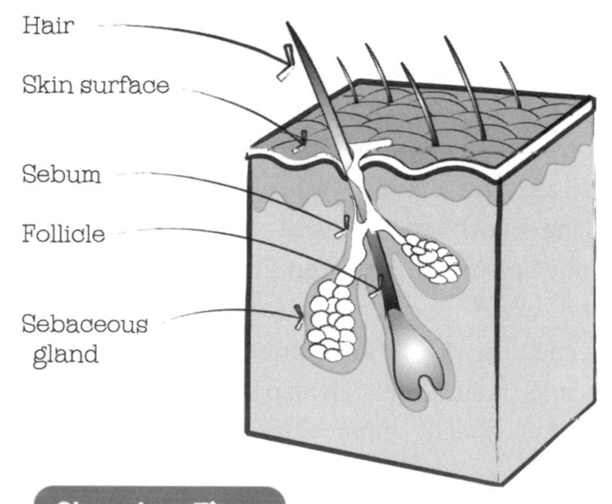

Sebaceous glands open into hair follicles, which in turn open onto the skin.

The glands produce an oily secretion called sebum.

Sebum is a delightful mixture of fats, cholesterol and waxes.

Chocolate Zits

A possibility for the supposed link between chocolate and acne is that both chocolate and sebum are rich in fats.

sebaceous gland opens into a hair follicle, which in turn opens onto the skin. Sebaceous glands, which cover most of the body, are most abundant on the face and chest, but totally absent on the soles of feet and the palms of the hands. These glands produce an oily secretion called sebum — hence the name, sebaceous gland. It's normal for the body to produce sebum — a complex mixture of fats, such as triglycerides, cholesterol and waxes.

In acne, the sebaceous gland is overactive. A mix of sebum, keratin and pieces of dead cells plugs up the mouth of the gland, which then balloons up and swells.

Sometimes the acne gets worse. In these more complicated cases of acne, bacteria invade and infect the distended gland. Even worse, the contents of the swollen gland can leak into the surrounding tissue, causing inflammation and a cyst.

Why did people try to make a link between chocolate and acne? The answer is that both chocolate and sebum are rich in fats. People assumed that if you ate lots of chocolate, your body would then try to get rid of the fat in the chocolate by squirting it out of your sebaceous glands. The theory was that the overloaded sebaceous glands would then get blocked, and degenerate into acne.

Although there was absolutely no proof for this theory, dermatology books in the 1950s blandly claimed that chocolate caused acne. From the 1960s to the 1980s, dermatologists tried to prove a link between eating chocolate and getting acne.

A few clever studies tracked some of the fats in chocolate using very low levels of radiation, to see where the fats ended up. But hardly any of the tracked fats in the chocolate appeared in the sebum, proving that there was no molecular link between chocolate and acne.

Another study got a group of male prisoners and a group of teenagers to eat two types of 'chocolate' bars that tasted almost the same. Half of the 'chocolate' bars contained 10 times the regular amount of chocolate, while the other half had absolutely no chocolate in them at all. This study showed that chocolate did not cause any increase in acne.

In plain English, studies completed so far show that chocolate does not cause acne. This is good news!

And coming back to delicious chocolate and the old saying that 'you are what you eat', why on Earth would you object to being sweet, desired and universally tempting? After all, the cacao tree from which chocolate is made is called *Theobroma* — which means 'food of the gods'.

Squeeze a Pimple?

No, never. If you squeeze a pimple, you will physically damage the hair follicle. You will also set off cellular reactions, which lead to an increased release of leukotrienes and enzymes, which lead to inflammation and scarring. The scar is a flat, pigmented brownish scar — it will fade, but only after several years.

If you can't stand the look of the swollen head of a pimple full of pus, or if it is painful because it is so full of pus, there is something you can do. First, sterilise a needle or a sharp pin with methylated spirits or a flame. Second (and get someone else with a very steady hand to do this), pierce the head of the pimple, but only pierce the surface (to no deeper than a millimetre). Then, wipe away the pus with a clean cloth or paper tissue (it doesn't have to be sterile). Whatever you do, do not squeeze the pimple.

On the other hand, some people do get immense joy from squeezing a pimple ...

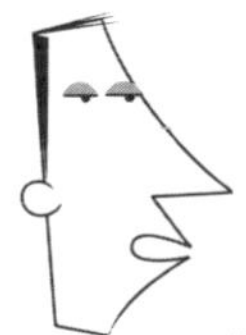

References

Cains, G.D., 'Acne vulgaris', *MIMS Disease Index*, Intercontinental Medical Statistics (Australasia) Pty Ltd., 1991/92, pp. 15–17.

De Launey, Wallace E., *Principle and practice of dermatology*, Butterworths Pty Ltd, 1978, pp. 84–89.

Baby Delivery Myths

For most of recorded history, female midwives were in charge of the medical care of pregnant women. A major change occurred in the 17th century, when male European physicians began delivering the babies of the nobility. By the early 20th century, medical science was being applied to delivering babies. Yet even today there are still many myths about how to speed up the delivery of an overdue baby. Most of these beliefs have been proven not to work, and many of them can be risky.

(On the other hand, induction of labour — trying to force the delivery of a baby by medical staff in a hospital — can be risky as well. Just because a procedure is done in a hospital does not mean that it's always safe, and just because it's done outside a hospital doesn't mean it's unsafe.)

First, some background knowledge. Generally, most women go into labour on their own, usually driven by signals from the baby. (For women having their first baby the average time for the beginning stages of labour is about 14 hours.) A labour can also be longer if the cervix is scarred from a previous delivery, if the contractions are weak, or if the baby is not properly lined up. Various chemicals or procedures are sometimes used as induction agents to speed up delivery — if the need arises.

One of the most common beliefs is that walking and exercise will bring on the delivery of the baby. Surprisingly, according to John Schaffir, of the Ohio State University College of Medicine and

Public Health, no studies have been done that follow women through their entire pregnancy, looking at the effects of exercise and walking. (Apparently, walking around does make some very pregnant women feel better — and that has to be good.) However, one study did find that very fit women have slightly shorter gestations — but no study has looked at average women. And indeed, while gentle exercise during pregnancy is fine, strenuous exercise is linked to smaller babies, preterm deliveries and pregnancy-induced hypertension.

Not one study has looked at sexual intercourse as a way to induce contractions (although many couples have tried this anyway). This method is believable, because both semen and standard induction agents contain chemicals called prostaglandins. Maternal orgasms late in pregnancy have also been linked to increased contractions.

Trials have shown that nipple stimulation for three hours per day for three days can help 'ripen the cervix'. This means that the

Baby ... where's the baby?

Baby Delivery

There are still many myths about how you can speed up the delivery of a baby. Most of these beliefs have been proven not to work, and many of them can be risky. On the other hand, induction of labour can be risky as well.

cervix is more ready to deliver the baby, sometimes triggering labour. But nipple stimulation is not always safe. It can be so powerful that it can stimulate the uterus into excess activity. There are even rare cases where nipple stimulation contracted the uterus so powerfully, that it squashed the baby, dropping the baby's heart rate to dangerously low levels. One such case needed an emergency Caesarean section. And one Australian obstetrician recently reported a case where a patient's breasts were so sore from nipple stimulation, that she was initially unable to breast-feed.

Laxatives are an old folklore favourite for initiating labour. They are moderately effective — but again, there are risks. Normally the baby has its first bowel movement *after* birth. But there are cases of laxatives being linked to babies having their first bowel movement while still inside the uterus. This contaminates their lungs as they float in the amniotic fluid, and can cause major problems once they try to breathe air.

Other folk remedies for inducing birth include various herbs (e.g. red raspberry, or blue or black cohosh). But red raspberry, while it makes contractions more regular, also makes them weaker. Enemas have long been a part of the preparation for labour — but there is no evidence that they help kick-start it. Indeed there have been cases of maternal and foetal death due to enemas. Finally, with regard to bringing on labour either by eating spicy food or frightening the mother, there are no reports on how effective, or risky these methods are.

There is one important lesson. Pregnancy is not a disease, and in most cases, both the mother and the baby will be well and healthy after the delivery — with no need for extra intervention.

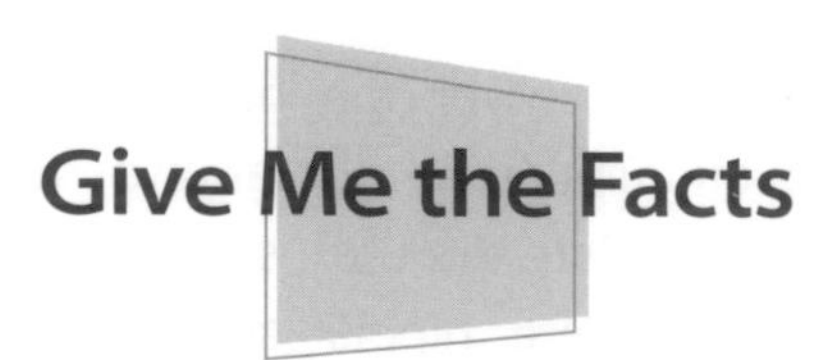

Give Me the Facts

A house built of stones is more than a pile of stones. In the same way, science is built on facts, but it is more than a pile of facts. Sir Karl Popper, the British philosopher of natural and social sciences wrote: 'An assertion is true if it corresponds to, or agrees with, the facts.'

But what if the facts change? After all, medicine is not a science, it's an art.

One study looked at 260 summaries which had been published between 1935 and 1995, in the journal *Surgery, Gynecology and Obstetrics*. They found that only half of the 'facts' were still regarded as 'true' 45 years later.

Another study looked at 474 conclusions in the field of liver disease, over the period 1945–1999, which had been published in the journals *Lancet* and *Gastroenterology*. Again, only half of the 'facts' were still regarded as 'true' 45 years later.

The British neurologist John Hughlings Jackson, was very wise when he said, 'It takes 50 years to get a wrong idea out of medicine, and 100 years a right one into medicine.'

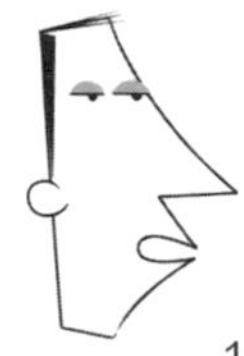

References

Blau, J.N., 'Half-life of truth in medicine', *The Lancet*, vol. 351, 31 January 1998, p. 376.

Hall, John C. & Platell, Cameron, 'Half-life of truth in surgical literature', *The Lancet*, vol. 350, 13 December 1997, p. 1752.

Schaffir, Jonathan, MD, 'Survey of folk beliefs about induction of labor', *BIRTH*, vol. 29, no. 1, March 2002, pp. 47–51.

Poynard, Thierry, MD, *et. al.*, 'Truth survival in clinical research: an evidence-based requiem?' *Annals of Internal Medicine*, vol. 136, no. 12, 18 June 2002, pp. 888–895.

Uluru to You

To most non-Australians, Uluru (previously known as Ayers Rock) is the ultimate symbol of Australia. It also has enormous cultural significance to the local Anangu Aborigines. The story told to the tourists, and to Australian schoolkids, is that Uluru is the largest monolith in the world. A monolith is a 'single stone', so Uluru is supposedly a giant pebble either sitting on, or partly buried in, the desert sands. However, geologists tell us that this is a mythconception.

The Aboriginal people had known Uluru for tens of thousands of years. The Europeans discovered it only recently. The explorer Ernest Giles sighted it from a great distance on 13 October 1872, but could not get closer, because he was blocked by Lake Amadeus. The explorer W.C. Gosse approached Uluru on 19 July 1873. He wrote: 'The hill as I approached, presented a most peculiar appearance, the upper portion being covered with holes or caves. When I got clear of the sandhills, and was only two miles distant, and the hill for the first time, coming fairly into view, what was my astonishment to find it was one immense rock rising abruptly from the plain; the holes I had noticed were caused by the water in some places forming immense caves.' Uluru is indeed impressive, rising over 300 m above the surrounding desert sands, with a circumference greater than 8 km.

Gosse gave Uluru the name of 'Ayers Rock', after Sir Henry Ayers, the Premier of South Australia at that time.

The idea of the giant pebble

It was thought that this monster rock had very little going on down below.

There's a lot going on down there

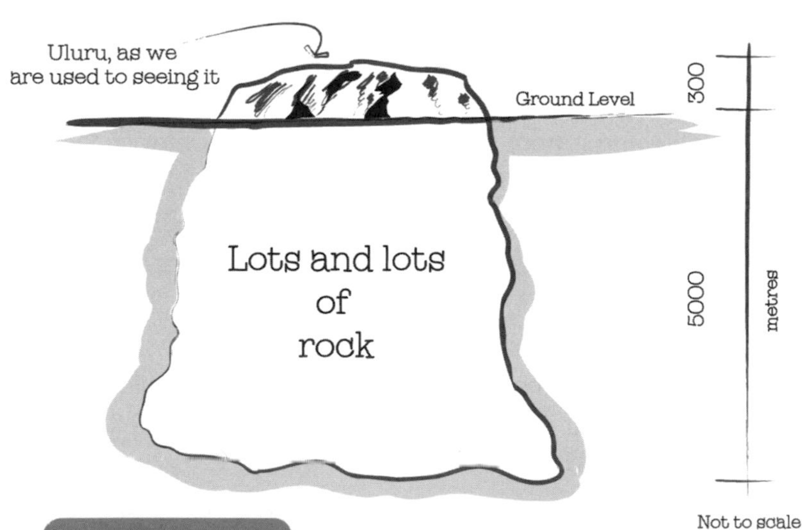

Uluru to You

It has long been thought that Uluru is the largest monolith (single stone) in the world. So, Uluru is supposedly a giant pebble sitting on, or partly buried in, the desert. Geologists tell us that this is very wrong.

From 900 to 600 million years ago, much of what is today Central Australia lay below sea level, in what is called the Amadeus Basin. Rivers dumped sand and gravel onto this sea floor until about 550 million years ago, when parts of the Amadeus Basin began to uplift. Uluru is made of this sand and gravel. Over a period of 100 million years (from around 300–400 million years ago), the land mass containing the future Uluru collided with other continents — leading to folding, faulting and more uplift. This slow-motion collision squeezed the sand and gravel sediments into rock — and also tipped them on their side through nearly 90°. The raised areas then began to erode away for the next few hundred million years.

A few locations (Uluru, Kata Tjuta and Mount Connor) survived above the surrounding areas, because they were made of harder rock that happened to have been cemented together with quartz. Around 65 million years ago, the local climate became a lot wetter. Rivers ran in the area, and sediments filled up the valleys between Uluru, Kata Tjuta and Mount Connor, smoothing out the landscape. Not much has happened (geologically speaking) over the past 65 million years except a bit more erosion.

So, is Uluru the biggest monolith?

First, Uluru is not the biggest Australian rock rising above the local flat plains. Mount Augustus in Western Australia is bigger.

Second, Uluru is not a giant isolated boulder, partly buried in the desert sand. Instead, it is part of a huge mostly underground rock formation, that is 100 km or so wide, and perhaps 5 km thick. The only three sections visible above the ground are Uluru, the magnificent domes of Kata Tjuta (formerly known as the Olgas) and the forgotten mountain, Mount Connor. The geologists, Sweet and Crick write that: 'Uluru is not a giant boulder, as one popular view would have us believe. The huge vertical "slab" of rock, of which Uluru is but the exposed tip, extends far below the surrounding plain ...'

On the other hand, the phrase 'largest monolith in the world' does make a nice story to entice the tourists. Just don't say it near a geologist.

Colour of Uluru

If you were to scrub the rock at Uluru with a brush and some soap, you would see that the base colour of the rock is grey. You can see this grey colour in some of the caves.

The red colour of Uluru is caused by rust, which is just iron oxide. In the past, there were huge mountains of iron oxide in the Australian outback. They weathered over hundreds of millions of years. The red dust from these mountains blew through the outback, colouring the entire landscape, including Uluru.

The colours of Uluru change at sunset because the sun is at such a low angle. The rays of the sun have to travel through a greater thickness of atmosphere. The blue light is bent away, leaving behind the red light — which makes Uluru glow radiantly red.

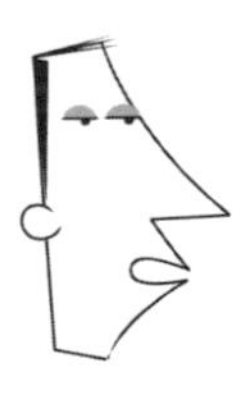

Reference

Sweet, I.P. & Crick, I.H., 'Uluru & Kata Tjutu: A Geological History', Australian Geological Survey Organisation, AGPS, Canberra, Australia, 1994.

Typhoid Mary

Typhoid fever is a very nasty disease, despite the availability of antibiotics such as chloramphenicol, which came onto the scene in 1948. Today, it still infects some 17 million people each year (mainly in Africa, South America and the southern and eastern regions of Asia), killing about 600 000 of them. Before the advent of antibiotics, typhoid was very much feared. This is partly why Mary Mallon was vilified and demonised as 'Typhoid Mary' in New York in the early 1900s. It was claimed that she deliberately and malevolently infected with typhoid everyone she came in contact with.

Today, the phrase 'Typhoid Mary' refers to someone who carries death, doom and destruction with them. It would be fair to imagine that Typhoid Mary killed thousands of people. But she didn't.

We have known the disease of typhoid for a long time. Hippocrates described it about 2400 years ago. Its name comes from the Greek word *typhos* meaning 'mist' or 'cloud', and refers to the confused state of mind of the sufferer. Typhoid appears to be the disease that killed Alexander the Great in 325 BC.

The disease is caused by a bacterium called *Salmonella typhi*, which lives only in humans. It is transmitted by the faecal-oral pathway, which is part of the reason why you should wash your hands after going to the toilet, and before preparing food. The bacterium also travels in the water supply, which is the major pathway of infection. Today, the disease is most common in areas of overcrowding, such as refugee camps and cities in poorer countries. The death rate varies between 1% and 20%, depending on whether antibiotics are available. About 1–2% of untreated sufferers become long-term carriers. Sometimes, the initial attack can be so mild that it is not recognised. This can lead to the 'symptomless carrier', who can infect others, and yet not believe that they themselves have ever been infected.

On 28 November 1902, one of the founders of the science of bacteriology, Robert Koch, introduced the concept of the 'healthy carrier of typhoid fever' to a scientific meeting in Berlin. He had gradually formed this idea over several years of close observation. Other European bacteriologists soon confirmed his findings. But the concept had definitely not yet entered the public consciousness.

This explains why, in the early 1900s, Mary Mallon refused to believe that she was a carrier. She was born in Ireland in 1869,

Mary, Mary . . . needs your lovin!

The Queen of Death and Destruction, Mary Mallon

Typhoid Mary

Typhoid fever is a very nasty disease and before the advent of antibiotics, it was even nastier. This is why Mary Mallon was vilified and demonised as 'Typhoid Mary' in New York in the early 1900s. They claimed she deliberately infected thousands.

and came to the United States in her teens. She was certainly not ignorant — she had a good writing hand, and loved to read Dickens and the *New York Times*. She was fiercely independent, tall, blonde, blue-eyed, and had a 'determined mouth and jaw'. She was also poor, female and Catholic, in a town that was run by wealthy male Protestants. Cooking was her passion and pleasure. She was so good at cooking, that when the wealthy family for whom she worked left Manhattan and took their summer holiday in a rented house on Long Island in 1906, they took her with them.

But the summer turned very unpleasant, as six of the 11 people in the household came down with typhoid fever. The owner of the house was worried that he would have difficulty renting the house in the future, if it were associated with typhoid. So he hired a New York sanitary engineer George A. Soper, to work out where the typhoid had come from. Soper soon excluded the drinking water as a possible source. He went looking further, and found that not only had typhoid hit seven of the eight houses in which Mary Mallon had been the cook, but also that one person had died.

Mary's very popular speciality was peach ice cream. Ice cream is an ideal incubator for *Salmonella typhi*, because it is both rich in fat (good food for the bacteria) and not cooked (so that the bacteria don't die). Soper told Mary his suspicions. According to a later version of the event, she refused to believe his accusation, and tried to attack him with a carving fork. He fled, returned with reinforcements, and had her incarcerated at Riverside Hospital for Communicable Diseases on remote North Brother Island, overlooking the Bronx. If this seems like an extreme response, you have to realise that in that year of 1906, 600 of the 3000 people who were infected with typhoid fever died.

After three years and a few lawsuits, Mary Mallon gained her freedom, by promising that she would no longer work as a cook. In 1910, she accepted the restraint, was released, and vanished into the teeming hordes of New York City.

Five years later, there was yet another typhoid outbreak of 25 cases with two deaths, this time at the Sloane Hospital for Women in Manhattan. Mary Mallon was found working in the kitchen as a cook, under the name of Mary Brown. This time, she was incarcerated even more quickly, and remained on North Brother Island for about 25 years until she died in 1938.

'Typhoid Mary' became known as a reckless woman who knowingly put others at risk. However, she was not the only 'symptomless carrier'. Several hundred other symptomless carriers were discovered in New York during the period of her incarceration. And although she was written up as a culinary 'grim reaper', and given the catchy nickname of 'Typhoid Mary', she was responsible for fewer than 50 cases of typhoid, and of those only three died. The newspapers of the day called carriers of typhoid 'living storehouses and factories of disease' and 'human culture tubes'.

In those days, before the existence of ethics committees who could weigh up public good against private rights, it was easy to make a harsh example of one woman, and suggest that she had killed thousands, rather than provide true statistics of deaths and infection rates.

Bacteria Attack

Bacteria are truly microscopic — much too small to be seen with the naked eye. So how do they attack us?

There are three main methods.

First, they can make an 'exotoxin', a poisonous chemical released from their tiny bodies. The bacterium *Clostridium botulinum*, famous for making the muscle paralyser (and wrinkle remover) called Botox, works this way.

Second, they can make an 'endotoxin'. This nasty chemical stays within their bodies until they die and their bodies disintegrate. *Salmonella typhi* does this.

The third method of attack does not involve a toxin at all. Instead, we human beings become sensitive to certain parts of the body of the bacterium. The bacterium *Mycobacterium tuberculosis*, that causes tuberculosis (commonly known as TB), attacks us this way.

References

Brooks, Janet, 'The sad and tragic life of Typhoid Mary', *Canadian Medical Association Journal*, 15 March 1996, pp. 915–916.

Chin, James, 'Typhoid fever', *Control of Communicable Diseases Manual*, American Public Health Association, 17th edition, Washington, DC, 2000, pp. 535–540.

Finkbeiner, Ann K., 'Quite Contrary', *The Sciences*, September/October, 1996, pp. 38–43.

Hasian Jr, Marouf A., 'Power, medical knowledge, and the rhetorical invention of 'Typhoid Mary'', *Journal of Medical Humanities*, vol. 21, no. 3, 2000, pp. 123–139.

Mikkelson, Barbara, 'Typhoid Mary', Urban legends homepage: www.snopes.com/medical/disease/typhoid.htm

One-way Mirror

If you watch enough detective movies, you will eventually see the 'one-way mirror' scene — the Bad Guy in a well-lit room, and the Good Guy in another well-lit room next door separated by the famous one-way mirror. (Note that *both* rooms are well lit.) We are supposed to believe that the magic one-way mirror lets the light go in only one direction, so that the Good Guys can see the Bad Guy but all the Bad Guy sees is a mirror.

Of course, this is rubbish! There is no such thing as a mirror that allows light to travel in only one direction. The fourth edition of the *American Heritage Dictionary of the English Language* believes in this myth. It defines a one-way mirror as 'a mirror that is reflective on one side and transparent on the other, often used in surveillance'. A one-way mirror is an impossibility.

However, if you juggle the light levels in the two rooms, you can fake the effect of a one-way mirror.

According to physicists, a mirror is any polished surface that reflects a ray of light. From Greco-Roman times to the European Middle Ages, a mirror was just a piece of metal (e.g. silver, bronze or tin) that had been highly polished. The next stage, coating glass with a thin layer of metal, began in the late 12th century. Typically, the metal, an amalgam of mercury and tin, was poured on, and then off again, leaving a thin coating of the shiny metal on the glass. The advantage of pouring the metal onto glass is that glass is easy to make flat, so that you get a nice even reflection. In

Ahhh ... the old one-way mirror trick

The secret revealed

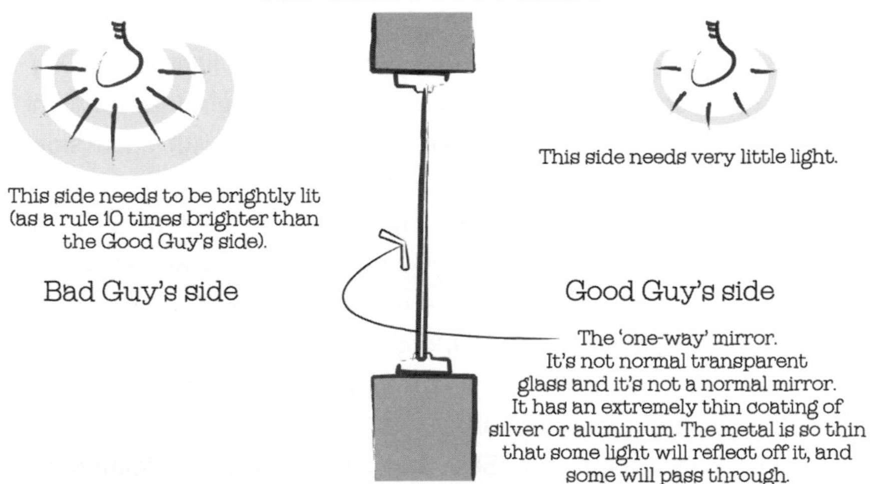

One-way Mirror

There is no such thing as a mirror that allows light to travel in only one direction. However, if you juggle the light levels in two adjacent rooms, you can fake the effect of a one-way mirror.

1835, Justus von Liebig worked out how to coat glass chemically (not mechanically) with silver. Today, we make mirrors by blasting a thin layer of silver or aluminium onto glass. Usually, there is an opaque layer under the silver or the aluminium.

You can understand how the so-called one-way mirror works if you imagine two rooms next to each other, and think of sound. In one room there are two people listening to very loud music, while in the other room there are two people whispering to each other. The whisperers can hear the loud music next door but the people in the room with the loud music cannot hear the whisperers.

The standard one-way mirror set-up (not just confined to movies) does the same thing with light, not sound.

First, you install a 'special' sheet of glass between the two rooms. It's not normal transparent glass and it's also not a normal mirror. It's called a half-silvered mirror. It has an extremely thin (much thinner than in a normal mirror) coating of silver or aluminium. The metal is so thin that while some light will reflect off it, some will pass through the metal. The normal opaque layer (that usually sits between the metal and the glass) is also missing.

This mirror will *reflect* a certain percentage of light (say 80%) from each direction, and will *pass* the rest of the light (say 20%) in each direction. It is just a regular two-way mirror with each side giving you the same amount of mirroring (80%).

Second, you fiddle with the light levels in the rooms. It is extremely important that the Bad Guy's room is very well lit, while the Good Guy's room is dark. As a rough rule, the lit room should be 10 times brighter than the dark room.

These two factors (half-silvered mirror, different light levels) will fake the effect of a one-way mirror

In the bright room, the Bad Guy sees a bright reflection of himself in the mirror. (We've all seen this when we try to peer through the window of a darkened shop on a sunny day.) There is also a very faint image of the Good Guy coming through the glass but this is massively washed out by his own bright Bad Guy reflection.

In the dark room, the Good Guy easily sees the Bad Guy in the well-lit room. The mirror will reflect the Good Guy's image but this is so faint, that it is washed out by the much brighter image of the Bad Guy.

So why do the movies always show the Good Guy in a well-lit room? Because some actors get paid about $40 million a movie, approximately $500,000 a minute. The film studio wants to give the audience value for money, so they light up the Good Guy.

One-way Mirror Impossible

A mirror that lets light through in only one direction is an impossibility.

Suppose that you made a six-sided box entirely out of this magical mirror material. Light would flow into the box, but could not get out again. You would end up with lots of light inside the box. This light would have some energy. You would have a magic box that would, by itself, fill up with energy. If you put a hole into the box, you could let the energy out and use it. Then you could block up the hole, and the box would magically fill up with energy again without having to do any work.

This box would create energy out of nothing, which goes against the Laws of Thermodynamics. A one-way mirror would allow you to build a perpetual motion machine.

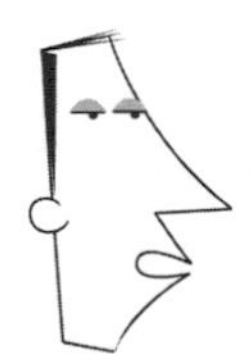

Reference

American Heritage Dictionary of the English Language, 4th edition, 2000.

CD-ROT

The oldest known musical instruments are a set of tuned bones, about 20 000 years old, made from the shoulder blades, hips, jaws, tusks and skullbones of mammoths. A team of criminologists, forensic scientists and musicologists deduced that these bones were primitive percussion instruments such as drums and cymbals. In fact, they still maintain their tone to the present day. Some people make a similar claim about the strength and durability of compact discs — but like the song in the Gershwin opera, *Porgy and Bess*, 'It ain't necessarily so'.

For most of its history, music was as temporary and ethereal as the wind. If you wanted to hear live music, you needed the services of live musicians.

This began to change around 1877, when Thomas Edison invented the 'phonograph', which recorded music as tiny dents in a sheet of tinfoil wrapped into a cylinder. The technology improved with Emil Berliner's invention of 1887 — which now recorded the sound as tiny bumps in a spiral groove on a flat black disc. The sound quality improved slowly as these gramophone records improved with different sizes of discs, thinner grooves to squeeze in more music, and different rates of spinning the records. The next big jump in sound quality came in 1958, with the introduction of stereophonic sound. There were now two channels, each with a different sound, feeding two separate speakers.

This record technology became almost obsolete in the 1980s, with the introduction of the compact disc. The belief was that they were as tough as nails, and would probably outlast the Pyramids. Popular TV science shows had presenters rubbing CDs in the dirt, and even drilling small holes in them — and afterwards, they still played perfectly. We were all assured that here was a truly archival medium — one in which our stored memories would live forever. But once again, we were deceived.

Forget CD-ROM and think CD-ROT.

Sony and Philips designed the original audio CD, which carry 74 minutes of music. This duration was chosen because Norio Ohga of Sony (who had studied opera in Berlin) decreed that a single CD should be able to carry all of Beethoven's Ninth Symphony (which plays for 70 minutes). And the reason for their particular size is that they are just a little too large to fit into a shirt pocket, making them just that much harder to steal.

CDs versus the cockroach

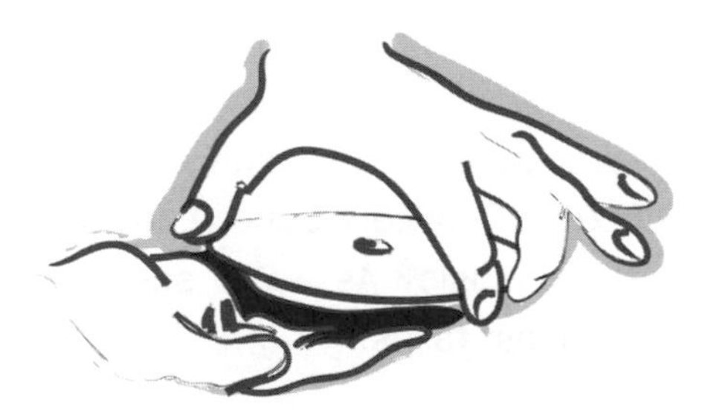

The supposedly robust CD
A CD can easily be damaged by scratches, sunlight, solvents, dirt and dust, or even by touching its surface.

The highly robust cockroach
Not easily damaged by scratches, sunlight, solvents, dirt and dust, touching, or nuclear weapons.

Compact Discs

The compact disc came into existence in the 1980s. It was believed that they were as tough as nails. TV science shows had presenters rubbing CDs in dirt and using them as coasters. Once again, we were deceived.

It didn't take long to squash the idea of compact discs being indestructible.

In the early days of CDs, there were problems with sulphur from cheap cardboard sleeves, and from some of the inks used to print the label information. The chemicals ate through the lacquer, destroying the thin aluminium layer. These teething problems were quickly recognised and fixed.

Today's CDs can be ruined by severe bending, or by rolling a sharp point (such as a ballpoint pen or a pencil) over the surface. Compact discs are also damaged by exposing them to sunlight or solvents, by peeling off the label, by dirt or dust, or even by touching the surface. (A handy hint: never buy a CD with a fingerprint on it — always ask for a clean one. The oils on fingertips can sometimes eat into plastics.)

Under archival storage conditions (low temperature and humidity) Kodak says that CDs will last for between 70 and 200 years. But who has archival conditions, with constant low temperature, low humidity and low light? In the average office, one of these expensive archival CDs will last only 100 years. Indeed, the heat of a car cabin in summer can decrease their life to as little as five years.

And yet the original *Domesday Book*, compiled in 1086 by William the Conqueror for accounting purposes, is still in mint condition in the Public Record Office in Chancery Lane, London. Does this mean that taxes will outlast music? As we've seen, it took a mammoth effort for music to outlast taxes.

Microscopic CD

A compact disc is a 1.2 mm-thick disc of transparent polycarbonate (as used in bulletproof glass), with a diameter of about 12 cm. It stores the information in a five-kilometre-long spiral of raised rectangular bumps on the surface. These bumps are about 0.5 microns wide (about 100 times smaller than the thickness of a human hair) and one-eighth of a micron high. This makes the CD one of the smallest easily available, mechanically manufactured objects ever made by human beings.

Because it is virtually impossible to see tiny transparent bumps on a transparent disc a thin layer of metal (usually aluminium, although gold and silver have been used) is laid on top of the polycarbonate. This shiny metal reflects the laser beam, so that it can read the little bumps. Unfortunately, this metal layer is very fragile, and is easily damaged. For this reason the metal layer is then covered with a thin layer of lacquer to protect it from the air, as well as dirt and chemical attack. The title of the CD is usually screen-printed onto this lacquer.

References

Fox, Barry, 'Can CD companies stop the rot?', *New Scientist*, no. 1902, 4 December 1993, p. 19.

Hillenbrand, Barry, 'The sumo halls are alive', *Time*, 9 March 1992, p. 56.

21 Grams

The trailer for the 2003 movie, *21 Grams*, begins with quite a compelling but totally false statement: 'They say that we all lose 21 grams at the exact moment of death.' It's a short and sweet attention-grabber — but the science behind the statement is non-existent.

For hundreds, possibly thousands, of years people have believed that the 'soul' has a definite physical presence. But it was not until 1907, that a certain Dr Duncan MacDougall of Haverhill in Massachusetts actually tried to weigh the soul. For his study he recruited six terminally ill patients. In his office, he had a special bed 'arranged on a light framework built upon very delicately balanced platform beam scales'. He claimed that the scales were accurate to two-tenths of an ounce (around 5.6 g). Knowing that a dying person might thrash around and upset such a delicate set of scales, he decided to 'select a patient dying with a disease that produces great exhaustion, the death occurring with little or no muscular movement, because in such a case, the beam could be kept more perfectly at balance and any loss occurring readily noted'.

He published his paper, called 'The Soul: hypothesis containing soul substance together with experimental evidence of the existence of such substance', in the April 1907 edition of the journal *American Medicine*. He asserted that he had measured a weight loss at the moment of death, which he claimed was associated with the soul leaving the body. He wrote, from beside the special bed of one of his patients, that 'at the end of three hours and 40 minutes he expired and suddenly coincident with death the beam end dropped with an audible stroke hitting against the lower limiting bar and remaining there with no rebound. The loss was ascertained to be three-fourths of an ounce'.

He was even more encouraged when he repeated his experiment with 15 dogs, which registered no change in weight in their moment of death. This fitted in perfectly with a popular belief that a dog had no soul, and therefore would register no loss of weight at the moment of demise.

But even before his article appeared in *American Medicine*, the *New York Times* — on 11 March 1907 — published a story on him, entitled 'Soul Has Weight, Physician Thinks'. His reputation was now assured, having been published in both a reputable medical journal as well as the *New York Times*.

As a result, the 'fact' that the soul weighed three-quarters of an ounce (roughly 21 grams) became common knowledge.

This weight will be the death of me

21 Grams

The 'soul' is reported to have a physical mass of approximately 21 grams or three-fourths of an ounce. If this is true (which hasn't been proven) how can 21 grams float?

But when you actually look closely at his scientific work, some problems become apparent.

First, a group of six dying patients is not a large enough sample size. Any student of basic statistics knows that the results from such a sample would not be statistically significant.

Second, and most damningly, he got the results he wanted (i.e. the patient irreversibly lost weight at the moment of death) from only *one* of the six patients! Two results were excluded because of 'technical difficulties'. Another patient only temporarily lost the three-fourths of an ounce — and then put the weight back on! (Did his soul came back?) Two of the other patients registered an immediate loss of weight at the moment of death, but then lost more weight a few minutes later. (Did they die twice?) So he didn't use six results, just one. This is a very good example of 'selective reporting'. He kept the one result that propped up his pet theory and ignored the five results that didn't.

The third problem is a little more subtle. Even today, with all of our sophisticated technology, it is still sometimes very difficult to determine the precise moment of death. And which death did he mean — cellular death, brain death, physical death, heart death, legal death? How could Dr MacDougall be so precise in 1907? And how accurate and precise were his scales?

From this single result that could not be reproduced, an enduring myth was born. There may indeed be lightness after death — but this experiment didn't prove it.

We do leave something behind us when we die — the enduring impact that we have had on others. Why not try measuring the impression of this mental impact, instead of attempting to measure the weight of the soul?

Soul Floats Up, or Drops Down?

There's something wrong with the idea of the body getting lighter as the soul leaves it. After all, if the soul is lighter than air, and floats upwards, then the body left behind should be left heavier.

Imagine that you are sitting, balanced on the end of a see-saw. You are tied to a few large helium balloons. These balloons want to rise — they have 'lift'. Indeed, these balloons make you lighter than your normal weight. You cut the string that ties you to the balloons. They immediately float upwards, away from you. They no longer give you any 'negative weight'. So you should get heavier, and the other end of the see-saw should go up.

But this didn't happen in the soul-bed experiment of Dr MacDougall. Instead, the other end of the balance went downhill.

This implies (if his experiment 'worked') that the soul is heavier than air. When it leaves the body, it should fall to the ground and flow downhill, rather than float lightly up to heaven. Perhaps if the soul is heavier than air (like a bird), it could stay aloft by flapping its wings ...

References

'Soul has weight physician thinks', *New York Times*, 11 March 1907, p. 5.

'Soul Man', www.snopes.com/religion/soulweight.asp: Urban legends reference pages.

MacDougall, Duncan, MD, 'The Soul: Hypothesis concerning soul substance together with experimental evidence of the existence of such substance', *American Medicine*, April 1907.

Cerebral Palsy and Birth

In the 1860s, the English surgeon, William Little became the first person to medically describe the condition known today as 'cerebral palsy'. Cerebral means 'related to the brain', while palsy means 'inability to fully control body movements'. Based on very little evidence, he suggested that a temporary lack of oxygen during the process of birth damaged certain parts of the brain, causing cerebral palsy. He was wrong, but even today, multi-million dollar compensation payouts are made based on this enduring myth.

The children that Little saw had stiff, frozen limbs. They had difficulty crawling, walking and grabbing objects. Strangely, as they grew older, the disease did not progress. Back then, the condition was called Little's Disease.

Cerebral palsy is not just one disease, but a group of many diseases with many different causes. It's a long-term disability of the central nervous system which causes poor control of posture and movement. It appears in the early stages of life, but is not related to any of the known progressive neurological diseases. It affects about two in every 1000 children, with about 500 000 victims in the United States alone. About 10% of children with cerebral palsy contract it after they are born, e.g., from meningitis, head injury, near-drowning, etc. The symptoms range from very mild to quite severe.

There are four main types of cerebral palsy. *Spastic cerebral palsy* (the most common type) affects about 70–80% of sufferers. The muscles become stiff and permanently contracted. The second type is *athetoid* (or dyskinetic) *cerebral palsy*, and involves uncontrolled, slow, writhing movements. It affects about 10–20% of sufferers. The third type is *ataxic cerebral palsy*, and it affects balance and depth coordination. It accounts for 5–10% of cerebral palsy sufferers. The fourth type is the mixed form, which has varying combinations of the first three types. The most common version is the combination of spasticity and athetoid movements.

About one-third of children with cerebral palsy have normal intelligence, one-third have mild intellectual impairment, while the remaining third have moderate or severe impairment. About one-half of children with cerebral palsy have seizures. They can also have delayed growth, and impaired vision or hearing.

In 1897, the psychiatrist Sigmund Freud, disagreed with the notion that lack of oxygen during birth caused cerebral palsy. He noticed that children with this disease often had other problems, such as seizures, mental retardation and disturbances of vision. He believed that the causes of cerebral palsy happened much earlier, while the foetus was still growing in the uterus. Freud wrote: 'Difficult birth, in certain cases, is merely a symptom of deeper effects that influence the development of the foetus.'

However, the general public still clung to the idea that cerebral palsy was caused by a temporary lack of oxygen during birth — and they still do today.

But that's not what the medical profession thinks.

In January 2003, after three years of deliberations, two separate groups of doctors (the American College of Obstetricians and Gynecologists and the American Academy of Pediatrics) released their publication called *Neonatal Encephalopathy and Cerebral Palsy: Defining the Pathogenesis and Pathophysiology*. The findings of this report were accepted by medical organisations worldwide. It confirmed yet again the opinion of obstetricians around the globe — that most causes of cerebral palsy (at least 90%) occur before the child is born.

Infections during pregnancy are probably the main cause. Other causes include chronic oxygen deprivation during pregnancy, developmental or metabolic abnormalities while in the uterus, auto-immune or coagulation defects in the mother, jaundice and Rh factor incompatibility in the newborn. And yes, a few per cent are probably caused by oxygen shortage in the brain or injury to the head during labour and delivery.

Giving birth is always a little risky.

Over the past 40 years we have vastly improved the chances of a safe delivery for both mother and child. During that time there has been no significant change in the incidence of cerebral palsy — despite the introduction of electronic foetal monitoring, massive increases in inductions and Caesarean sections, and enormous improvements in neonatal care.

Cerebral Palsy and Birth

Cerebral

(cer•e•bral) Of or relating to the brain or cerebrum

Palsy

(pal•sy) A weakening or debilitating influence

Cerebral Palsy

It's not just one disease, but a group of many diseases with many different causes. It's a long-term disability of the central nervous system which causes poor control of posture and movement ... and it appears in the early stages of life.

Babies who weigh less than 1500 g at birth make up a tiny percentage of newborn babies — but they make up 25% of all cases of cerebral palsy. This is a good indicator that the same factor that makes the baby small, also makes the baby more likely to suffer cerebral palsy.

Since cerebral palsy is often recognised at or just after birth, it led to the naive assumption that it was caused by the medical team — a good example of 'shooting the messenger'.

People with cerebral palsy do have special needs — which usually means they need money. In Australia, in 2004, the only way to obtain a significant amount of money is through litigation. Of the total obstetric malpractice payments, about 60% of the money goes to cerebral palsy cases. On average, legal costs eat up 70% of the money awarded. The result is that the person who needs the money gets less than half of the payout — a very wasteful process.

Wouldn't it be better to have a system where all the available money went to the people with cerebral palsy? Perhaps the lawyers could deliver the babies.

Risky Childbirth

Most people have normal births and normal babies. But sometimes things go wrong — and no one is at fault.

Consider the problem of the 'prolapsed cord'. The umbilical cord enters the vagina before the baby's head does. As the baby's head comes down, it squashes the cord, robbing the baby of oxygen. Even in a major teaching hospital, with all its modern facilities and emergency teams on stand-by, 42 minutes will elapse from the recognition of such an emergency, to the eventual delivery of the baby.

References

MacLennan, Alastair, 'A template for defining a causal relation between acute intrapartum events and cerebral palsy: international consensus statement', *British Medical Journal*, vol. 319, 16 October 1999, pp. 1054–1059.

Motluk, Alison, 'Inflammation may cause cerebral palsy', *New Scientist*, no. 2182, 17 April 1999, p. 21.

Seppa, N., 'Infections may underlie cerebral palsy', *Science News*, vol. 154, no. 16, 17 October 1998, p. 244.

American College of Obstetricians, Gynecology & American Academy of Pediatrics, 'Neonatal Encephalopathy and Cerebral Palsy: Defining the Pathogenesis and Pathophysiology', *ACOG Task Force Report*, USA, January 2003.

Thanks

Give it up for ... all my peeps in the HCPhood: MC Sandra Davies, Missy Alison Urquhart, Lil' Erin Young and Yo Judi Rowe.

My main man Adam Yazxhi for da tags ... the fully wicked Susan Skelly and da subbies at the *Good Weekend* Mag. Word up to Rose Creswell and Lesley McFadzean, Mixmaster Caroline Pegram and Ma Lady Mary D and The Home Crew.

'She does exist!'

Childhood geek with exceedingly high pants

Karl, the milkshake philosopher, 1968

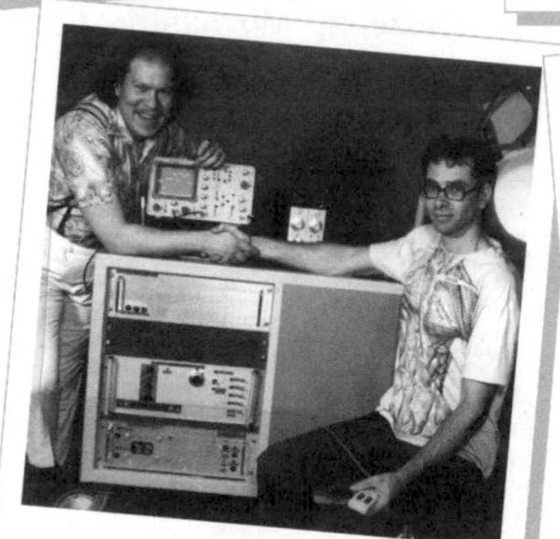

Jacqui and Karl, 1968

Karl in New York, 1978

It's a D-cup!

What avoidance behaviour can I do next?

Mel Bampton, Story Musgrave and Karl

Karl in Himalayas, 2004

Yak...Yak...Yak

Karl and Ian 'on-line' Allen

Pucker-up mike!

Da Home Crew

Ya! You can't get enuff orf Dr Karl!

Pigeon Poo, the Universe & Car Paint
and other awesome science moments
Dr
KARL KRUSZELNICKI'S
new moments in science #1

Flying Lasers, Robofish & cities of Slime
and more brain-bending science moments
Dr
KARL KRUSZELNICKI'S
new moments in science #2

Munching Maggots, Noah's Flood & TV Heart Attacks
and other cataclysmic science moments
Dr
KARL KRUSZELNICKI'S
new moments in science #3

Fidgeting Fat, Exploding Meat & Gobbling Whirly Birds
and other delicious science moments
Dr
KARL KRUSZELNICKI'S
new moments in science #4

Q & A with Dr.K
Why it is so
Headless Chickens, Bathroom Queues and Belly Button Blues
Karl Kruszelnicki

Dr Karl's Collection of
Great Australian Facts & Firsts
Great Australian facts & firsts
Dr Karl Kruszelnicki

Bumbreath Botox and Bubbles
and other Fully Sick Science Moments
Dr Karl Kruszelnicki